FARMING AND BIRDS

FARMING & BIRDS

RAYMOND J. O'CONNOR & MICHAEL SHRUBB

WASH DRAWINGS BY DONALD WATSON

CAMBRIDGE UNIVERSITY PRESS

Cambridge

New York Port Chester

Melbourne Sydney

Published by the Press Syndicate of the University of Cambridge
The Pitt Building, Trumpington Street, Cambridge CB2 1RP
40 West 20th Street, New York, NY 10011, USA
10 Stamford Road, Oakleigh, Melbourne 3166, Australia

First published 1986
First paperback edition 1990

Printed in Great Britain by the University Press, Cambridge

British Library cataloguing in publication data

O'Connor, Raymond J.
Farming and birds.
1. Birds – Great Britain 2. Agricultural pests
– Great Britain
I. Title II. Shrubb, Michael
598.2′52′22 QL690.G7

Library of Congress cataloguing in publication data

O'Connor, Raymond J.
Farming and birds.
Bibliography
Includes index.
1. Bird populations. 2. Birds–Habitat
3. Birds, Protection of. 4. Agricultural ecology.
I. Shrubb, Michael. II. Title.
QL677.4.026 1986 598.2′5264 86-2279

ISBN 0 521 32447 5 hard covers
ISBN 0 521 38973 9 paperback

This book is a product of a research programme commissioned by the Nature Conservancy Council and carried out by the British Trust for Ornithology

Contents

Preface

It is well known that farming can affect birds by altering the structure of farmland – by clearing woodland, removing or planting hedges, by filling in or creating ponds, and so on. Yet different ways of operating a farm can also influence its bird populations, even without any change in the farming landscape. The literature on farmland birds over the last three decades has been dominated by pesticides and habitat destruction (particularly loss of hedgerows) but agriculture has meanwhile become an industry dominated by technical innovation and resulting changes in management regimes, a factor which has been much underestimated in considering conservation on farmland.

Concern over pesticides generated an extensive programme of research into the environmental effects of agrochemicals and eventually gave rise to the Pesticides Safety Precautions Scheme. In this, the agrochemicals industry accepts independent scrutiny to ensure that there will be no possibility of environmental hazard in newly developed chemicals. Despite (or perhaps because of) its voluntary nature, the Scheme has been successful in minimising incidents of mass deaths and of breeding failure among birds, such as characterised the original introduction of persistent organochlorine seed-dressings. Any incidents detected are routinely investigated by MAFF or DAFS as appropriate and a programme of monitoring and research in respect of toxic chemical residues and their effects on the environment is maintained by the Nature Conservancy Council (NCC). Nevertheless, the long-term ecological implications of pesticide use, particularly the influence they might have on the selection of farming systems, remain relatively unstudied. The development of knowledge about habitat conservation for farmland birds has also been relatively poor. Because only limited funds are available to it, the Nature Conservancy Council has sought within agriculture mainly to minimise further loss of the remaining semi-natural habitats in agricultural land and to monitor wildlife trends there, for example through its support for the Common Birds Census (see below). In ornithological circles, however, imagination was seized by the question of hedgerow destruction. Emphasis on this has probably been excessive and has been compounded by so much of the available empirical data coming from rather few (and at times conflicting) case studies.

Our book attempts to review the larger topic of the effects of agricultural development on bird populations in Britain. We have excluded such specialist 'farming' enterprises as fish farming and deer ranching from consideration. Within these limits we review relationships between farm structure and birds but our primary concern is to identify the changing effects on birds of modern agricultural management. Britain's agricultural policy is primarily (and until very recently was solely) addressed to increased production. This has resulted in subsidies being available to bring into production areas other government agencies are charged with keeping out of production. These policy anomalies mostly relate to patches of semi-natural habitat on farmland and are outside our remit. Our thesis is that technology tends to outstrip policy in ways that have gone largely unconsidered and that this process then affects significant fractions of farmland bird populations in unexpected ways. A good example of this is the widely accepted idea that pastoral and dairy farming are favourable, and cereal farming unfavourable, to birds. Our evidence convinces us that it is mixed stock and crop farming that provides the best agricultural environment for birds and that the practices of intensive dairy farming are as damaging to bird communities as are those of specialist cereal farms.

Our emphasis is on providing new evidence as to how modern agricultural practices affect farmland bird populations. Consequently, this is not a book on the ecology of farmland birds. Such a book is needed – indeed, in our researches we were struck by just how little is known about how birds use farmland. Many of the points we raise are new to or little emphasised in the existing literature and we have no doubt that the ideas we put forward here will be modified by further study and experience. Nevertheless, we were greatly encouraged by the results of a survey we undertook of the views of the long-term participants in the BTO's Common Birds Census, for we found that many of the features we detect at the population level have also been apparent to those experienced in the censusing of farmland birds.

We are writing here for those interested in or needing factual information on the fortunes of bird populations on farmland in Britain. Our data analyses have necessarily been statistical but, as our findings are intended for a wide audience, we have avoided the presentation of formal statistics; these we intend to publish in the technical literature. Our access to the massive files of the British Trust for Ornithology has allowed us to use the entire databank of farmland Common Birds Census results since its inception in 1962, as well as to use the nest record cards for farmland species over the same period. The CBC was originally started, at the request of the NCC (then the Nature Conservancy), in the light of concern about the effects of toxic chemicals but it now places greater emphasis on assessing the effects of changing agricultural practices. We relate the results from the CBC and nest record schemes to data from MAFF's June Census statistics and Pesticide Usage Surveys, and from Rothamsted Experimental Station's Surveys of Fertiliser Practice. We have also had access to the results of a number of ancillary schemes and projects, some of them as yet unpublished, which yielded data on key points which would otherwise have gone uncovered. One of our problems has in fact been the wealth of data available to us: so much information resulted that we were forced to adopt a cut-off date (1980) for various analyses. Further data have accrued in the meantime and such is the speed of agricultural development that in some cases these additional data already suggest effects on birds not apparent in the earlier material. We have also drawn heavily on observations over the past 25 years at Oakhurst, the Shrubbs' family farm comprising 427 acres (173 ha) of cereals and grass at Sidlesham in south-west Sussex, and have drawn on the experience of long-term CBC workers of the changes on the farms they have studied. We found it invidious to acknowledge the use of individual data from these workers since the same ideas often came from several people; instead, we have listed below all those who contributed information.

Chapter 1 provides some historical background to agriculture, against which the changes in farming and bird populations must be judged. Chapters 2, 3 and 4 consider the composition of farmland bird communities and relevant aspects of breeding and feeding ecology. Chapter 5 examines the post-war growth of agricultural specialisation and its general implications for birds. Chapter 6 looks at the bird habitats of farmland, but hedgerows are considered separately in Chapter 7 in view of previous emphasis and controversy on this subject. Chapter 8 is a major review of the specific impacts of farm management on birds, though earlier chapters discuss management effects where appropriate. Chapter 9 looks rather briefly at the pattern of agrochemical usage, with emphasis on some less familiar aspects of the topic, before considering in Chapter 10 the effects of crop protection and game management on farmland species. A final short chapter draws together a number of themes common to some of the earlier chapters.

We are grateful to many people for a variety of help and assistance afforded us in the writing of this book. Most of the agricultural statistics used here were extracted in the MAFF Library in London and the assistance of the staff there is gratefully acknowledged. We also thank David Donaldson (Chichester MAFF) and Ann Griffiths (West Sussex County Council) for their help in this area. The volume of analysis of BTO data undertaken would have been impossible without computer use. The computer files used were input over several years by BTO technical staff working on projects funded by the Natural Environment Research Council, Manpower Services Commission, MAFF and EURING. We are grateful to these staff for their efforts and to the organisations mentioned for their finance. The present staff of the Trust gave much help and advice in using these data, particularly Stephen Baillie, Robert Fuller, David Glue, John Marchant, Robert Morgan, Kenny Taylor, and Phil Whittington. We also benefited from the use of unpublished analyses by Leo Batten, Robert Fuller, Gillian Griffiths, Peter Lack, and Chris Mead. Data preparation expressly for this book was provided by Mercedes Gibbs, Elizabeth McHugh, and Susan Miller. We also thank Robert Fuller, Chris Mead, and Jim Wolf for taking on most of Raymond O'Connor's administrative duties during the final writing of the manuscript.

We would like to thank Ron MacDonald, Patrick Osborne, Robert Prŷs-Jones, and Hugh Robertson for use of material first published in their doctoral theses. James Cadbury, Eric Carter, Rhys Green, Peter Greig-Smith, Dick Potts, Michael Rands, Ken Smith, Iain Taylor, Stephanie Tyler, and Mike Wareing kindly provided us with data or information from their own and their colleagues' work. We are grateful to those who supplied photographs or allowed us to reproduce diagrams, as acknowledged individually in the captions. Our first draft of this book was extensively reviewed by several friends and colleagues – Robert Fuller, Rhys Green, Frank Gribble, Peter Lack, Mike Pienkowski, Dick Potts, and Chris Whittles – and we appreciate the time and effort they put into this task. For secretarial help at various stages we thank Audrey Causer, Dorothy Rushton, Janet Mewis, and especially Elizabeth Murray who both provided secretarial support to Raymond O'Connor and prepared the majority of the diagrams for the book. We also thank Kenny Taylor for preparing prints of the photographs used here, and Donald Watson for his splendid illustrations. Finally, we thank our wives Deirdre and Veronica for their patience through what turned out to be a bigger task than any of us foresaw.

Raymond O'Connor's post at the BTO is funded by the Nature Conservancy Council who also support the Common Birds Census and Nest Records Scheme. This support is gratefully acknowledged.

Finally, we cannot conclude the acknowledgements of a book such as this without expressing appreciation of the many hours of fieldwork put into the CBC and Nest Records Scheme by BTO members; without their labours, this book would not have been possible. We cannot name them all here – there are over a thousand for the CBC alone – but we would like to thank by name the following who responded to our request for information on their experience of the effects on birds of management and habitat changes on their CBC plots: G. Atkin, M. Archdale, E. R. Austin, G. C. Avison, E. Bagshaw, B. & A. Bailey, M. Bailey, T. Banyard, H. J. C. Bashford, R. C. Branwhite, A. Bull, J. M. Butterworth, L. Charlton, A. R. Crawford, O. Dansie, H. Dean, P. J. Edwards, I. English, J. Field, R. Fox, B. Gillam, B. Gilling, P. Gordon, A. C. Gutteridge, H. W. Hamar, A. Henderson, J. A. Hopwood, R. Irvine, W. S. Jacklin, P. Jennings, H. U. Lennox, A. Mandell, J. McCowen, J. Milson, L. J. Milton, M. Newton, L. Nottage, M. F. Oliver, F. Parsons, D. Penman, L. W. Pygott, A. J. Prater, G. Redfern, G. Riddle, C. E. V. Saxton, D. Scott, E. M. P. Scott, J. T. R. Sharrock, T. Simpson, E. & M. Smith, J. Stevenson, D. Summers-Smith, R. J. Taylor, E. Ward, M. Wareing, B. M. Wherrett, S. Willoughby, G. O. Wilson, and E. J. Wiseman.

1: The historical background to modern agriculture

Pressure on the land, such as is experienced today, is nothing new. In terms of the agricultural technology available, the land surface of lowland Britain has been under intense pressure from the human population, probably since Roman times and certainly since the Norman Conquest, Stamp (1955). A simple way to appreciate this is to look at stocking rates. Stamp (1955) estimated that in England and Wales around 1340 the national herd was some 1.6 million cattle (half of which were plough teams) and some ten million sheep. To support this number of animals today would require some 1.60 million hectares (14.7 per cent of the total land available) of grassland of all types. For the mid-nineteenth century, at the peak of Victorian high farming, the figure would have been *c.* 2.83 million hectares (22 per cent of the total land available) and for medieval England and Wales, when far less was known about the improvement of pasture and when arable crops were fed to stock on a more limited scale, it seems reasonable to double the area to *c.* 5.66 million hectares. A similar area, of which between one-third and one-half was fallow annually, was required for arable crops, for yields were no more than a tenth of today's levels.

With such pressures on the land, farming systems produced some of our more distinctive habitats. Lowland heathlands, chalk downland, upland moorland, water meadows and hedgerows are, at best, semi-natural habitats formed either by grazing pressures (the first three) or management systems (the remaining two). Inevitably, major changes in farming systems and techniques produced major ecological changes in the countryside. Farming history in Britain has many examples but Moore (1965) singles out three as having had radical effects on farmland birds. He lists, first, the original change from forests to primitive farms; second, the change from the medieval open-field system to enclosed farms; and third, the change from traditional use of horse and manpower to the use of the internal combustion engine and chemicals. We have little information on the first and second, though we do know that the introduction of the heavy Saxon plough, with its team of up to eight oxen, brought a change in field shape and size for the same reason that increased tractor power has done so recently. Higher power needs larger and more conveniently shaped working areas to give the most economic result, since turning wastes time and energy.

In medieval farming the development of the three-field system increased cropped land by 33 per cent. Open-field systems and farms composed of individually enclosed fields both existed but only the latter allowed mixed farming of a more modern type. Open-field agriculture was largely based on a simple rotation system. Each village had five or six blocks of land: one devoted to houses and their gardens, two devoted to grazing, and two or three devoted to cultivation. One of the grazing blocks was meadowland for hay and winter keep for stock, the other was for common grazing. The arable blocks were tilled either as winter corn and fallow (the two-field system) or as winter corn, spring corn and fallow (the three-field system), and the fallow land was used for grazing, which manured it. Many stock were slaughtered each autumn to minimise the pressure on winter keep before the introduction of turnips and clover for this purpose in the eighteenth century. However, the three-field system itself meant increased cereal production, some of which was fed to stock, which in turn led to a rise in stocking rates. Because the animals were either tethered or attended by cowherds or shepherds there was little need for stock-proofed enclosure of the fields and hedges were essentially absent. Some idea of the avifauna of such lands can be obtained from a survey of the parish of Laxton in Nottinghamshire, one of the few parishes in lowland England that has never been enclosed (Pollard, Hooper & Moore, 1974). Farmland there remains under the medieval three-field system and lacks trees or bushes, and most of the commoner hedgerow birds are absent. The community is instead dominated by Skylark, Partridges (both species), and Lapwing. Such results lend emphasis to the great importance of hedgerow and small woodlands in supporting the far richer songbird communities of modern farmland.

The wool boom of the later medieval period in Britain led to enclosure and the extensive conversion of arable to pasture. As Trow-Smith (1951) points out, this change was largely permitted because the Black Death had caused rural depopulation and a severe scarcity of labour. Arable yields were such that it is doubtful that the changeover would otherwise have occurred. The agricultural revolution of the eighteenth and nineteenth century brought further extensive enclosures, particularly in the Midlands, which included the subdivision of larger but already enclosed fields. As an example, Pollard *et al.* (1974) quote the already enclosed parish of Buckworth in Huntingdonshire, where field sizes declined from an average of *c.* 81 acres (32.4 ha) in 1680 to an average of *c.* 16 acres (6.5 ha) in 1839. In the process, *c.* 17 acres (6.88 ha) of land was taken out of production through creation of an additional 28 000 yards (25.6 km) of hedge. The reduction in field size presumably resulted from the interaction of rotational mixed farming and a relatively small average holding size. Even as late as 1875 some 82 per cent of all holdings in England and Wales were of under 100 acres (40 ha). The need for adequate field drainage may have been involved in these changes too. For example, a comparison of the 1846 Tithe Maps for the parish of Sidlesham in Sussex with the 1876 Ordnance Survey shows extensive hedge and ditch clearance to have occurred between the two surveys, producing an increase in field size from an average of 6.9 acres (2.8 ha) to an average of 8.9 acres (3.6 ha). This change coincided with the introduction of mass-produced clay tile drains for field drainage in the 1840s. The lines of old ditches (and, indeed, the existing ones) suggest very strongly that a close network of open ditches was previously relied upon to get surface water away from crops and that this consideration was more important than convenient shapes for working or the loss of land. It is interesting to note that the different examples of Buckworth and Sidlesham included either the creation or removal of some notably awkwardly shaped fields.

As well as enclosure by hedgerows, the farming devised in the eighteenth and nineteenth centuries brought another important ecological change. The classic four-course Norfolk rotation – of clover for manuring (the compost crop), wheat, roots (the cleaning crop) and barley – resulted in a great increase in actual tillage. The complete rotation probably produced a 50 per cent increase (compared with the three-field system) in the use of the plough and other tilling machines and also introduced inter-row cultivation by hoeing, an operation unknown to the medieval farmer. It was thus essentially a system suited for light or easily worked and free-draining soils. Fertility was maintained by the compost and cleaning crops and by extra manuring as a result of the much higher stocking rates permitted by the introduction of root crops for winter keep. These soils were unsuited to the arable methods of medieval farmers, so the new methods resulted in pronounced tendencies for arable to shift to areas of light soils, formerly perhaps largely waste (i.e. rough grazing), and for pasture to shift to heavy soils, particularly in parts of the Midlands (Smith, 1949).

We can only speculate on the impact of this change on farmland birds since historical accounts of bird distribution and numbers generally lack the statistical data needed. A few glimpses of major shifts in distribution are afforded us. Thus the Great Bustard was driven off the Wolds of Yorkshire as cereals took over there and as the windbreaks planted to shelter the crops broke up the open plains of the bustard's habitat. Similarly, the Stone Curlew was apparently common on the Brecks in the first half of the nineteenth century but decreased markedly in the 1840s and 1850s as the new agriculture (and forestry) consumed 28 per cent of the area (Newton, 1896). The species recovered there at the end of the century when the area reverted to heathland as a result of the agricultural depression (Ticehurst, 1932). Conversely, the nineteenth century expansion of the range of the Stock Dove in Britain was in response to the spread of arable crops (Murton, 1971).

All periods of agricultural change thus far considered have a common theme. They involved changes in the arable/pasture balance, nationally and regionally, and in the shape and size of working areas (fields) in response to technical innovation. This theme has continued in the modern agricultural revolution but in more recent years economics have been increasingly important in deciding the form and direction of farming. The impact of economics cuts right across the patterns discussed above; they affect every farm irrespective of its natural status.

The 1870s marked the last high-water mark of the agricultural revolution started in the mid-eighteenth century. From about 1875 the doctrine of free trade and the increasing flood of cheap high-quality American wheat onto English markets, made possible by major improvements in ocean transport, initiated a long decline culminating in the Depression of the 1920s and 1930s. The effects of this on farming and therefore on the countryside are set out in Tables 1.1–1.3 for the years 1873–75 and 1930–32. Table 1.1 summarises the area of the three major farming land uses – pasture, rough grazing, and arable – and of the three major crop types – cereals, temporary grassland or leys, and roots crops – as a percenage of the total land area available. The area of tilled land in Britain had fallen to 8.47 million acres (3.43 million ha) by the early 1930s. This was significantly less than existed in England and Wales in 1698, when Gregory King estimated 9 million acres (3.64 million ha) of arable (by which he meant tilled land) (Hoskins & Stamp, 1963). It may, in fact, have been lower then than at any time since the medieval period for a population of around 40 million, compared with 5.2 million in 1698 and 3–4 million in the period before the Black Death. Eighty-one per cent of the total agricultural area was in grass by 1930–32, of which 46 per cent was classified as pasture, 43 per cent as rough grazing and the rest temporary grass. These bare figures say little of the real status of this grassland but the probability is that a significant proportion was arable land simply allowed to revert, rather than properly established pasture. A serious decline in the quality of grassland is indicated by the figures for stock in Table 1.2. By 1930–32 the total herd, expressed as grazing units, had increased by 2 per cent but overall stocking rates had declined by 12 per cent. The extent of the decline varied regionally. For example, in eastern England it was as high as 48 per cent whilst in Wales it was only 3 per cent. In general, the south and east were more affected than the north and west.

The Wryneck provides one of the few documented examples of the effects on birds of the deterioration of grassland during the Depression in Britain (Peal, 1968). Wrynecks are predominantly confined to areas of chalk and greensand (though historically other records of birds on heavy clay are available) and feed extensively on ants. Peal attributed a decline in Wrynecks between 1939 and 1966 to the extension of arable farming and forestry over that period but he also noted that over the previous 70 years (except 1914–18) Wrynecks were already becoming scarcer in England. Since the area of arable land and woodland was steadily decreasing (and grassland increasing) throughout this time (see above), some quality of the grassland or its management and its suitability for ant populations is implicated (Peal, 1968). In neglected ungrazed areas the nursery mounds of the ant *Lasius flavus* are small, perhaps only a quarter of the volume of those in grazed areas, since successful colony foundation is inhibited by shading by tall herbage (Potts, 1970). This effect may account for the

Table 1.1. *Areas (million ha) of farmland and crops in Britain 1873–1983. Percentages are of total land area*

	Pasture		Rough grass[a]		Arable		Cereals		Ley grass		Roots	
	Area	%[b]	Area	%[b]	Area	%[b]	Area	%[b]	Area	%[b]	Area	%[b]
1873–75												
England	4.15	32.1	1.16	****	5.47	42.3	3.02	23.3	1.05	8.1	1.17	9.0
Wales	0.71	34.4	0.57	****	0.45	21.8	0.23	11.1	0.15	7.2	0.06	2.9
Scotland	0.44	5.5	4.31	****	1.40	17.7	0.57	7.2	0.54	6.8	0.27	3.4
1930–32												
England	5.49	42.6	1.82	14.1	3.54	28.1	1.56	12.1	0.88	6.8	0.85	6.6
Wales	0.85	41.2	0.87	42.2	0.25	12.1	0.09	4.3	0.11	5.3	0.03	1.4
Scotland	0.63	8.1	4.47	50.2	1.23	15.9	0.40	5.2	0.60	7.7	0.17	2.2
1961–63												
England	3.9	27.8	1.31	10.1	5.22	40.5	2.54	19.7	1.62	12.5	0.82	6.3
Wales	0.70	33.9	0.66	32.0	0.34	16.5	0.08	3.9	0.21	10.2	0.04	1.9
Scotland	0.35	4.5	5.02	65.0	1.38	17.8	0.41	5.3	0.76	9.8	0.18	2.3
1981–83												
England	3.18	25.1	1.01	8.0	5.27	41.7	3.34	26.4	0.98	7.7	0.63	5.0
Wales	0.84	40.5	0.53	25.6	0.26	12.5	0.06	2.9	0.16	7.7	0.02	0.9
Scotland	0.56	7.3	4.26	47.7	1.12	14.5	0.52	6.7	0.49	6.3	0.09	1.2

[a] Rough grass figures for the first period are for 1892, the first year that figures were collected. Commons grazings are included.
[b] Expressed as a percentage of total land.
Data from MAFF/DAFS June Census statistics (see Appendix 3).

decline recorded for Wrynecks, as it may for the decline in grassland Grey Partridge populations (where ants may constitute over 30 per cent of the chick diet). As we will see in Chapter 5, though, a variety of other species actually benefited and spread through Britain in the wake of agricultural neglect.

The financial pressures which caused the farming slump are indicated in Table 1.3. Faced with a decline of up to 50 per cent in the purchasing value of crops, farmers survived by spending little or nothing. Even so, between 1916 and 1940 some 2.92 million acres (1.18 million ha) of crops and grass, excluding rough grazing, went out of use in England and Wales, 2.15 million (0.87 million ha) being lost from agriculture altogether (HMSO, 1968) at an average rate of *c.* 89000 acres (36000 ha) per year. This compares to an average annual loss of *c.* 30000 acres (12000 ha) since 1940, since when the human population has increased by 40 per cent. Of the land lost, *c.* 77 per cent went to urban and industrial use. Much of this was good land since population and settlement have historically been placed in the more fertile areas. The widespread distribution of many habitats in farmland, such as wet grassland or other unimproved pasture

Table 1.2. *Numbers of sheep and cattle and stocking rates in Britain, 1873–1983*

		Total cattle	Total sheep	Stocking rate[a] units/1000 ha
1873–75	England	4188689	19177160	*c.* 6300[b]
	Wales	696995	3205308	*c.* 4700[b]
	Scotland	1148661	7260467	*c.* 2800[b]
1930–32	England	5270118	13258276	4836
	Wales	820826	4261888	4571
	Scotland	1190811	7769590	2717
1961–63	England	7545217	13841050	7909
	Wales	1176220	5621866	7326
	Scotland	2016935	8629276	3052
1981–83	England	7962909	15370902	10674
	Wales	1412214	8380177	10092
	Scotland	2316440	7923399	4132

[a] Stocking rates expressed as grazing units per 1000 ha of grass of all types. One head of cattle = 5 grazing units; one sheep = 1 grazing unit.
[b] Area of rough grazing used in calculating stocking rates for this period was that for 1892, the first year for which rough grazings data are available.
Data from MAFF/DAFS June Census statistics (see Appendix 3).

and overgrown hedgerows, is the product therefore not of a traditional agriculture in tune with its environment, as is so often claimed, but of enforced neglect by an agriculture unable to maintain its capital. There is a certain irony in the similarity of some habitats thus created to the extensive fens and mires permanently lost to agricultural prosperity a century earlier.

Three points stand out from this brief survey of the Depression years. The unrestricted waste of farmland then has greatly increased subsequent pressure to improve marginal land and replace it, a continuing pattern only now beginning to be alleviated by policies of urban renewal rather than of urban sprawl. Secondly, the national herd has risen throughout this period: in times of depression because arable farmers retreat to stock rearing and in times of prosperity because stock farmers invest more. The increased herd has then to be accommodated on a smaller area of grass on each revival of cereal growing. The pressure on marginal grassland habitats is thus particularly severe in any upswing in agriculture's fortunes (Table 1.2). Thirdly, it was the situation reached in 1930–32 that was historically exceptional, rather than those of the 1870s and 1980s.

The position reached by agriculture in the early 1930s was untenable, both ecologically and economically, and Beresford (1975) has described the post-war revolution in farming as a renaissance rather than a revolution. To the extent that it has largely restored the 1875 position, this is true. But modern agriculture includes major new political and technical elements, which justify the term revolution. The three most important are the political and economic decision that a prosperous agriculture is necessary, the general replacement of muscle power (draught animals and men) by machines in agriculture and transport, and the steady and rapid

Table 1.3. *Yields, prices and value of wheat in Britain 1873–1983*

	Average yield tons/ha	Average price[a] £/ton	Average return £/ha	One hectare of wheat pays one man's wages[b] for:
1873–75	2.02	12.30	24.89	*c.* 34 weeks
1930–32	2.02	6.40	12.94	7–10 weeks
1961–63	3.87	21.11	81.69	7–10 weeks
1981–83	6.05	110.58	669.00	6–8 weeks

[a] Prices are for England and Wales only.
[b] Calculated on the average wage for basic hours.
Source: MAFF statistics.

development of agricultural chemicals. It is these features, and the speed with which they have been introduced under the continuing political and economic pressures on farmers, that have brought agriculture and conservation into conflict in the 1970s and 1980s (Hooper, 1984). Wildlife can adapt to change, and has done so in the past, but it cannot keep pace with the speed of modern technological development.

The main plank of Government agricultural policy after the 1939–45 war was the 1947 Agricultural Act. This set out a system of guaranteed prices for produce by acreage or deficiency payments, which made up the price obtained by the farmer in the market to a predetermined level, and created a web of capital grants to encourage and assist farm improvement and renovation. The levels of prices and grants were reviewed annually and often varied in response to changes in policy or perceived needs. The basic aim of the Act, however, clearly was to reconcile the long-standing policy of seeking cheap food supplies for the industrial working population both with the idea of maintaining a viable agriculture and with the markedly changed imperial and economic circumstances in which Britain found herself. In doing so, the Act extended and confirmed such pre-war decisions as the Wheat Act of 1932 (which first guaranteed wheat prices) and the active encouragement since 1925 of sugarbeet growing in East Anglia, intended to offset the decline in arable farming there. Although the instruments of policy have varied considerably since 1947, particularly with accession to the EEC, the essence of agricultural policy has remained unchanged until recently, and broadly the same route has been followed by most West European countries.

Summary

Land in lowland Britain has been under pressure from human population and agriculture since the early medieval period and this pressure has been instrumental in creating several major bird habitats. Historically, farming systems have changed in response to technical advances, and these changes have centred on the arable–pasture balance and the organisation of fields. The Depression of the 1920s and 1930s is the major historical exception to this general pattern of evolution in farming, and the post-war agricultural revolution has largely only restored agriculture to the position it held in 1875. The implementation of new technology, particularly in agrochemicals, has been crucial in this achievement.

2: The farmland bird community

Much of the British avifauna is of forest or woodland origin, dating from the time when the whole of Britain was dominated by woodland and by extensive low-lying marshes, and when there was only a small proportion of mountain tops and coastal habitat which provided open country. Over the past 5000 years many forest species have become adapted to using the remaining woodland edges, copses and plantations in open country, as the original forest was reclaimed by man for settlement and food production. After the Enclosure Acts of the eighteenth and nineteenth centuries such species found additional habitat in the network of newly created hedgerows, so much so that even contemporary writers remarked on the increase in songbirds in the countryside. Typical farmland countryside in England is therefore now dominated by a wide variety of hedges and attendant trees which Williamson (1967) estimated, on the basis of Common Birds Census maps, to amount to 12 km per square kilometre of countryside in England and Wales. In the Midlands and in the south of England much of this complex hedgerow system is little more than 200–300 years old (Moore, Hooper & Davis, 1967). This historical transition from forest to field system in lowland Britain explains why forest species account for about 80 per cent of farmland birds and why typical open country or 'steppe' species remain poorly represented. Such field species – particularly partridges, Lapwing, Skylark and Yellow Wagtail – together contribute only about 7–14 per cent of the birds of lowland farms in Britain (Williamson, 1967). Lowland farm communities in Britain are thus very different from those of North America and Africa, where much agricultural land has come from open plains and where species of open ground are more diverse and more numerous. In upland Britain bird communities are rather different from those of the lowlands, for land use there has been dominated by intensive sheep grazing accompanied by moorland burning, and stone walls and earth banks topped by gorse and bramble have taken place of hedgerow, particularly in the Celtic field systems of the west.

Moore (1980) defined farmland as excluding urban areas, large woodlands and plantations, large lakes and reservoirs, coastal areas, and uplands which are not grazed by sheep or cattle. Throughout this farmland small parcels of semi-natural habitat such as woods, copses, hedges, marsh and bog, are inextricably intermingled. Their frequency very often depends on the intensity and nature of the farm management and nearly all such parcels are managed to some extent. On this basis farmland is the largest bird habitat available and Moore estimated that within it there are between 50 and 100 million pairs of breeding birds of about 130 species out of the 230 or so that breed in Britain. Of these 130 species 81 (62 per cent) have small populations (below 100000 pairs) and are therefore vulnerable to habitat loss or pesticide mortality. In contrast, only 17 species have large populations (above 1 million pairs). The remaining 32 species have between 100000 and 1 million pairs. Other species frequently recorded on British farmland, such as Swift and House Martin, are not dependent on agricultural habitats as such and are only incidentally present in farmland censuses.

Williamson (1967) has suggested that a definition of a farmland bird community in the central English lowlands (after excluding pigeons and corvids) is 'a community in which Blackbird and Skylark occur roughly in the proportion of two, three or four to one, together making up between 15 and 40 per cent of the population'. Our own more extensive data show that this suggestion is valid for cereal and pastoral farms but in areas of intensive tillage Blackbirds and Skylarks may be equally numerous (see below). The Blackbird is the commonest species on agricultural land in the lowlands, though not in the highland zone (see below), and constitutes between 10 and 20 per cent of the population. Blackbirds are also predominant in parkland, in suburban gardens and in woodland, but Skylarks are largely absent from these habitats. Williamson (1967) found that a central Midlands farm sample varied from 100 pairs of 22 species to 499 pairs of 33 species; most plots had 24–35 species, and passerines accounted for over 95 per cent. The richest farms ornithologically were those with greatest fragmentation of habitat and containing not only hedgerows with tall trees, but streams or ditches with cover on the banks, small copses and shelter belts, pools, and farm dwellings with small orchards and gardens.

Fig. 2.1 summarises our own data on the abundance and distribution of the principal farmland species during the breeding season. For each species the figure plots average density on farmland against the percentage of Common Birds Census (CBC) farmland censuses in which the species was recorded. Species recorded in fewer than 20 per cent of the censuses are excluded here. There are a number of biases in this use of CBC data, particularly the under-representation of certain important farmland species (see Appendix 1), and densities and site frequencies also vary with

Fig. 2.1. The abundance of farmland birds in relation to the frequency with which they occur on CBC farmland plots 1962–81. Site frequency is the percentage of farms on which the species was recorded. Species on 50 per cent or more of the censuses are identified by name.

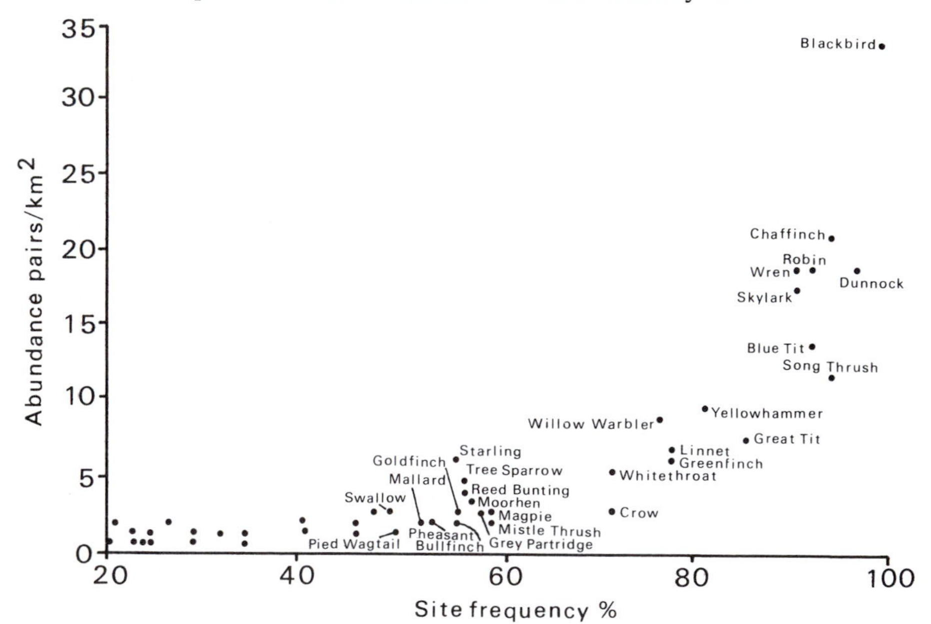

year-to-year changes in population level. Nevertheless, the basic structure of the farmland bird community, as revealed in Fig. 2.1, is quite robust. It comprises a relatively small group of widespread and abundant birds and a much larger group of less numerous species. The 15 most frequently recorded species in Fig. 2.1 are Blackbird, Chaffinch, Dunnock, Robin, Wren, Skylark, Blue Tit, Song Thrush, Yellowhammer, Willow Warbler, Great Tit, Linnet, Greenfinch, Starling and Whitethroat. Of these, only the Skylark is a field species and only the Willow Warbler and Whitethroat are summer migrants. Of the 11 most frequently recorded species here, all but the Skylark and Yellowhammer are also among the 11 most numerous species recorded in woodland by Fuller (1982). On any individual farm, these highly adaptable species are likely to be among the most numerous species present.

A considerable number of additional species occur in lower densities at some but not all farmland sites (Shrubb, 1970). The presence of these scarcer species of the community depends on the structure of the local habitat but some 14 of these species are nevertheless widely distributed, being found in half or more of the farmland sites. Examples of this group include Mallard, Moorhen, Swallow, Pied Wagtail, Bullfinch, and Tree Sparrow. For most species of this group, only one or two pairs are present on any one farm but a few, e.g. Tree Sparrow, may be more numerous where present. The rarest species are those that have particular habitat requirements only locally present as fragments of agricultural land. Species of this type include Reed and Sedge Warblers, Redshank, and Snipe. A few are strongly influenced by farming practice: the Corn Bunting, for example, is a bird of cultivations rather than of grassland.

Williamson regarded the Dunnock as perhaps the second most successful species on farmland, accounting for 8–26 per cent of the passerine total. It is a bird of woodland edge and of scrub rather than of forest proper and is therefore well suited to the linear scrub of hedgerows. Murton (1971) suggests that this species has probably colonised the western Palearctic only in comparatively recent times. Earlier it was largely confined to the scrub of forest edge and the tree-line in Poland and western Russia, extending above the tree-line into the gulleys of the broken montane habitat. It spread westwards as deforestation and scrub clearance for agriculture created an artificial approximation, in the form of ditches and hedgerows, to its native habitat. Murton noted its high densities in south-west England where double hedgerows associated with earthbanks and dark gulleys provided particularly close approximation to its ancestral habitats. Even in eastern England, however, Dunnocks retain something of this preference, for they are more numerous along hedgerows paralleled by ditches than they are in unditched hedges (Arnold, 1983). In contrast, Robin and Chaffinch are more restricted on farmland than are Blackbird and Dunnock, since they are more particularly 'forest' species. They require well-grown hedgerows with trees and as a result average only 6 per cent and 8 per cent of farmland communities.

Pigeons and corvids were omitted in Williamson's assessment of the farmland community and these species are also underrepresented in our data. The Woodpigeon – the commonest pigeon – is much more numerous as a winter or non-breeding visitor than as a nesting species. Amongst the corvids, though, the Rook is a species which perhaps many people regard as the typical farmland bird, in part because of its conspicuousness. The CBC, on which much of our quantitative review here is based, does not census the Rook adequately, but the species has been subject of two recent national surveys of its rookeries (Sage & Vernon, 1978; Sage & Whittington, 1985) which show it to be rather scarcer than its obvious behaviour must suggest to the casual observer. In terms of their relative consumption of energy, both Woodpigeon and Rook are more important members of the farmland community than their densities would suggest, being the largest of the commoner species on agricultural land. Murton (1971) estimated that Woodpigeon and Rook accounted for 40 per cent and 31 per cent respectively of the biomass of birds on his study plot at Carlton, Cam-

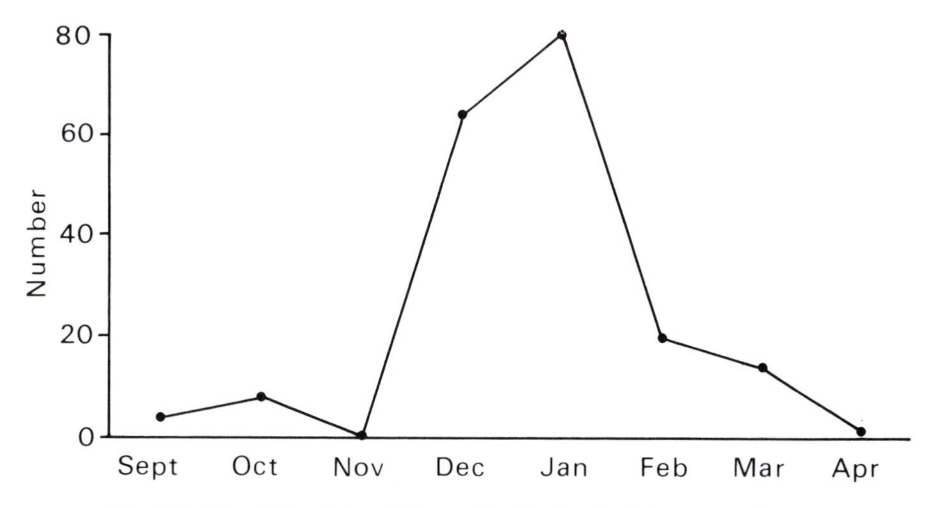

Fig. 2.2. The mid-winter increase in Starling numbers caught on Derbyshire farmland, 1973–84. Data *per* M. Wareing.

bridgeshire, although they formed only 14 per cent and 12 per cent of the bird density there.

Murton (1971) remarks on the absence of large numbers of summer visitors on farmland, due in part to the absence of a major spring flush of invertebrate food on farmland and in part because the resident species are highly efficient at exploiting what food there is available. Some recent evidence indicates that such summer visitors as do arrive on farmland are restricted to certain nesting habitats, from which they can expand only if the overwintering resident population has been badly depressed by severe weather (O'Connor, in prep.). In autumn, however, seed supplies increase seasonally, both from cereal grains and from weeds, and this increase accounts for the major presence of finch flocks on migration. Similarly, a winter peak of soil invertebrates supports winter visitors such as Lapwing, Starling, and thrushes. Fig. 2.2 shows this rise for Starlings at a regularly manned ringing site in Derbyshire.

Regional variation in agriculture

Farming in Britain is governed by soil, climate and relief which have always combined to impose a strong regional character in agriculture. In the past, this was strengthened by poor communications and by the self-contained nature of local communities, and even today there remains a strong streak of conservatism in farmers. Any map of land quality in Britain based on a grading system such as that used by MAFF (see below) has an immediately striking characteristic: most high-quality land is found in the east, where rainfall is lower, relief gentler, and soils less acid. This promotes the most striking division in British farming, the easterly distribution of cultivation and the westerly distribution of grass farming. The point is illustrated in Fig. 2.3. (This figure also shows the boundaries of the MAFF and DAFS regions as used in the June Census statistics from 1962 to 1980 and which we have adopted throughout this book unless otherwise stated.)

Nevertheless, the regional pattern is more complex. Thus, although the higher rainfall areas of the west favour grass rather than crops, areas of fertile soil in the far west have used their mild frost-free climate to specialise in early vegetables and flowers. The hill districts of Wales, northern England and Scotland are all stock-rearing areas, for no other form of farming is feasible. Areas of easy-working land near London and other major urban areas, particularly in the north-west under the rain shadow of the north Welsh mountains, have long grown market-garden crops. Modern transport facilities have helped these high input but valuable crops to become a major feature of the light silts and peats of eastern England, which are ideal for such crops as carrots or onions. Wheat traditionally succeeds best in a dry year and areas of lower rainfall are therefore favoured for the crop. Just over half (51 per cent) of the wheat in England and Wales is grown in the east and east Midlands, areas which average less than 30 inches (762 mm) of rain annually and comprise about 25 per cent of the total land available. Barley is the most widely grown cereal, reflecting its value as stock feed, but 31 per cent of the English and Welsh acreage is found in the chalkland counties of south and east England, whose light, often stony, alkaline soils are better suited to barley than to other cereals. In Scotland, oats, although recently displaced by barley, are the traditional cereal, as they succeed best in the damp, cool conditions and

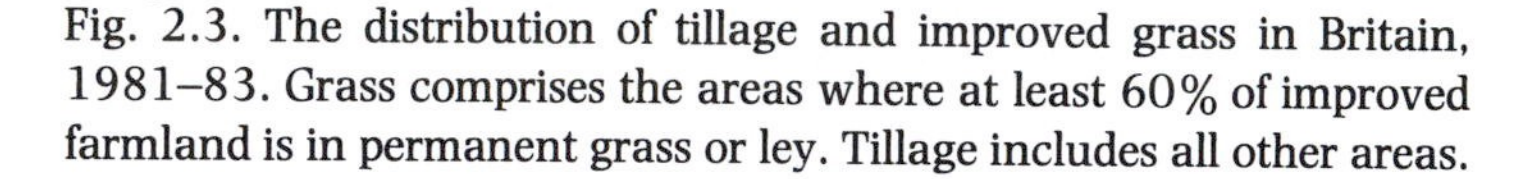
Fig. 2.3. The distribution of tillage and improved grass in Britain, 1981–83. Grass comprises the areas where at least 60% of improved farmland is in permanent grass or ley. Tillage includes all other areas.

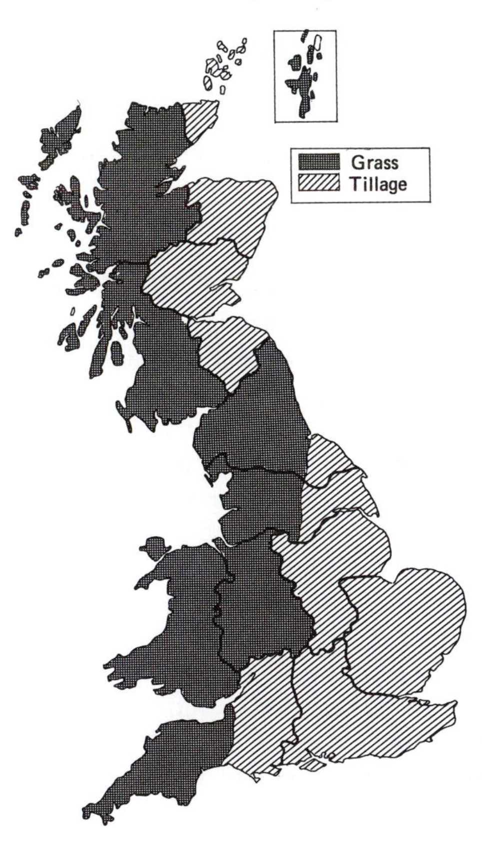

generally more acid soils found there. In England, however, oats have a strong southern distribution, which probably reflects their use as a break crop in intensive cereal rotations where roots are not grown. This difference is also reflected in the relative importance of Rook predation. In the south oats are predominantly winter-sown and Rooks are no serious problem since the wide variety of alternative food sources, e.g. stubbles, dilutes the impact of predation. Spring-sown oats are much more vulnerable, however, and their greater abundance (with other spring cereals) in the north underlies Scottish farmers' perception of the Rook as a pest species (Dunnet & Patterson, 1968; Feare, Dunnet & Patterson, 1974).

The best pasture is on soils of high fertility but which are often difficult to work; the great fattening pastures of the Midlands are on the heavy clay soils which formed the wheat belt of medieval England. Stocking rates are clear indicators of the fertility of grassland soils; these are highest in the Midlands, lowest in the north and in Wales and fall uniformly between these two extremes south of the Wash – Severn line. Here the marshes of Kent, Sussex, Norfolk and Somerset are important grassland areas whose fertile, but low-lying, soils have tended to be kept in grass, as arable crops suffer greater damage from flooding.

Broad though these patterns are, they are already reflected in the national distribution of some bird species. Thus, midwinter flocks of finches and buntings are a feature of extensive areas of arable farmland rather than of grassland, since the bulk of their food is derived from crop wastes and the weeds of cultivation. In Britain, therefore, more large flocks of wintering finches and buntings occur in the east and south rather than in the north and west, simply because there is a greater proportion of arable crops in these regions. Even on a local scale, the effects of this association may be seen, as in Sussex, where the bulk of finch flocks and arable land coincide along the Downs and the coast to the south (Shrubb, 1979), covering roughly 30 per cent of the county in a continuous band.

Some of the broad variations in farming practice just described arise for positive reasons and some for the negative reason that other, perhaps potentially more profitable, enterprises are barred because of adverse natural conditions. On a smaller scale any county or district shows the same variations for the same reasons, which makes for the historical diversity of British farmland and of its songbird populations. In every case they represent the best use of the land in the technological circumstances prevailing. One of the features of modern farming, however, is rapid technological development across its whole spectrum – in mechanical technology, in chemical control of crop environments, and in plant breeding. An important part of this development has been the speed with which it now is communicated to and adopted by farmers generally, which results in farmland becoming a progressively more unstable habitat. This process has been aggravated by economic instability.

Environmental factors

The three basic factors of soil, climate, and relief also strongly influence many aspects of farmland bird distribution. Fig. 2.4 shows the distribution of land quality in Britain. The Ministry of Agriculture's Land Classification of England and Wales divides farmland into five grades of excellence, from Grade 1 land, which has virtually no physical limitation to its use, to Grade 5 land which can be used largely only for grazing. The classification system

Fig. 2.4. Map of soil quality in England and Wales. Redrawn from Stamp (1955).

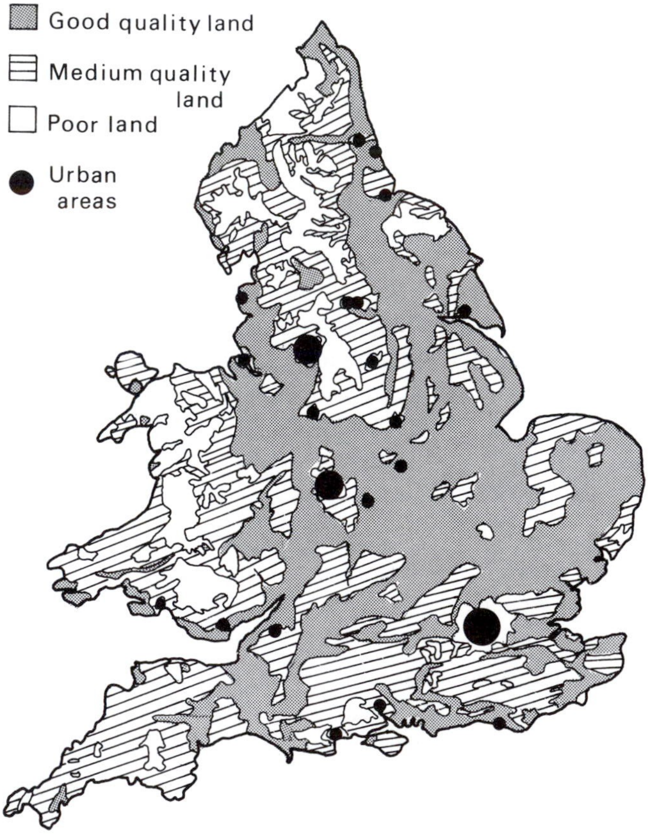

is based on the three natural factors just mentioned. Climate encompasses rainfall, transpiration, temperature and exposure; relief includes altitude, slope and surface irregularities; and soil characteristics include wetness, depth, texture, structure, stoniness, and available water capacity (Gilg, 1975). The approximate percentages of the total agricultural land in Britain and Wales comprising the different grades are 2.8 per cent in the Grade 1 category, 14.6 per cent in Grade 2, 48.9 per cent in Grade 3, and 33.7 per cent in Grades 4 and 5 (Beresford, 1975).

Soil quality is a fundamental factor in agriculture. The characteristics which typify the best soils are mechanical – they are naturally free draining and easily worked. Clearly the farmer can readily improve fertility by manuring or liming but poor drainage and, in particular, difficult soil structure are more limiting problems. Soil type and quality also limit the kinds and numbers of invertebrates adapted to live in them and this in turn limits the distribution of a number of bird species (Laursen, 1980). In Britain these include Wryneck, Lapwing and Sparrowhawk (Lister, 1964; Peal, 1968; Newton & Marquiss, 1976). In the case of Lapwings, some effect is certainly due to soil pH influencing the density of earthworms in the soil but the major effect on invertebrates is probably due to drainage. In summer Lapwing densities are highest on chalk and on sand and gravel, and are relatively scarce on clay soils; the well-drained soils are also preferred in winter. Golden Plover breeding densities also show a broad correlation with soil fertility (Ratcliffe, 1976). Drainage may influence the choice of nest sites: in Holland, for example, Lapwing nests were sited in meadows in relation to the depth of the watertable, except where this was more than 4 inches (10 cm) below the surface (Klomp, 1953). Ground-living species such as partridges and Lapwing are probably most susceptible to the influence of soil quality. In Denmark research has shown that only such species (and not arboreal ones such as titmice and warblers) were more abundant as the soil quality improved (Laursen, 1980). Curiously, though, Skylarks in Denmark showed a negative relationship with soil quality, being most numerous on the poorer soils, and in France the Skylark is the bird of grassland and the Crested Lark the bird of the richer arable fields, wherever the two species co-occur.

Climate is limiting in two ways, for whilst some crops succeed best in drier conditions, others need or tolerate a higher rainfall. The plant breeder can provide varieties which succeed in less than optimum conditions but the weather's impact on practical considerations of cultivation still limits what the farmer can do. Timeliness is all in farm operations. Williamson (1967) notes that in Britain climatic characteristics are subject to much more rapid change with altitude than in other temperate countries, so considerably greater variation in bird numbers is therefore to be expected in the highland zone of Britain than in the Midlands. The tree-line is unusually low in the British Isles, as compared with the rest of Europe, because of greater exposure to strong Atlantic winds and to the generally cooler and cloudier summers. On unimproved farmland the upper limit of settlement, i.e. habitation, is therefore generally just over 1500 feet (457 m), though sheep are grazed to higher altitudes. The length of the growing season (that part of the year during which mean temperature at the grass roots exceeds 6 °C, the minimum needed for active growth of crops) decreases sharply with altitude. Cumberland, for example, experiences a ten-week reduction in growing season at 1500–1800 feet (457–549 m) compared with sea level (Manley, 1952). In addition, average temperatures in spring fall sharply with altitude, especially in May, thus inhibiting the start of breeding by birds. Williamson suggested that bird populations at moderate elevations in the highland zone might be boosted by a population overspill from more suitable lowland haunts, e.g. whenever a succession of good breeding seasons or mild winters in the lowlands increases densities there. The Starling is one species whose populations in Scotland may vary in this way (O'Connor & Fuller, 1985).

Certain species seem to be especially susceptible to climatic effects. Thus,

Fig. 2.5. Number of breeding species on farms above (upper histogram) and below (lower histogram) 75 m, the median altitude of the 65 CBC farm plots considered. Data are for 1965.

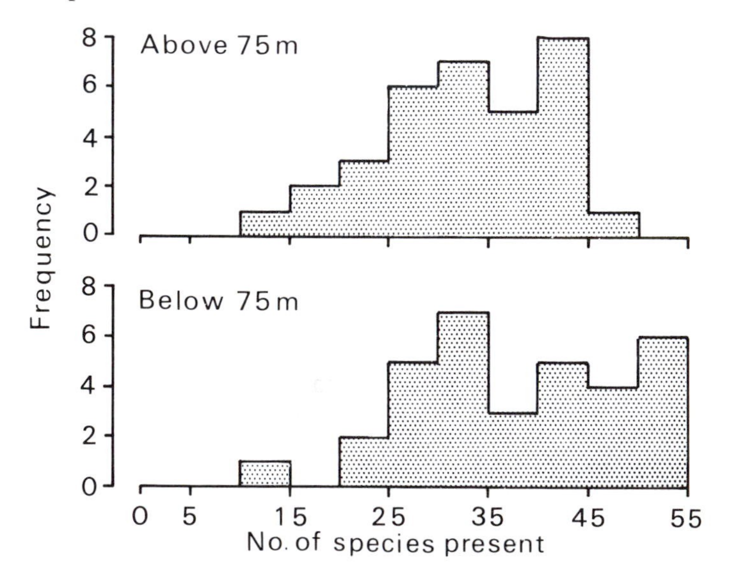

the Red-legged Partridge in East Anglia seems to be particularly favoured by continental weather. The decade 1940–50 was the first period since 1860 that the British climate became noticeably continental in nature, the winters getting colder but drier whilst the other seasons got warmer. According to Middleton & Huband (1965), the expansion of Red-legged Partridge populations in East Anglia appears to date from the fine anticyclonic summer of 1959. Again, Newton (1972) notes a general inability of Linnets and Goldfinches to cope with average British winters and this may explain why Linnets are an important constituent of winter finch flocks in the milder south.

The impact of relief is self-evident in agriculture and is sharply illustrated on the southern chalk, where the scarp slopes are too steep to plough and have largely gone out of agricultural use because farms there have specialised in cereal growing. But for birds also altitude and slope may matter. For some species it is the association of habitat with altitude that matters. Indeed, the very phrase 'upland birds' implies such association. Even for typically farmland birds, however, altitude may impose a limit, for farms below about 75 m are slightly richer in species complement than those at higher altitudes (Fig. 2.5).

Regional variation in bird communities

The composition of farmland bird communities varies markedly across the country, both in species complement and in the relative abundance of those species present. In general, species richness decreases from east to west (Fig. 2.6), without any great variation in the total densities of birds present,

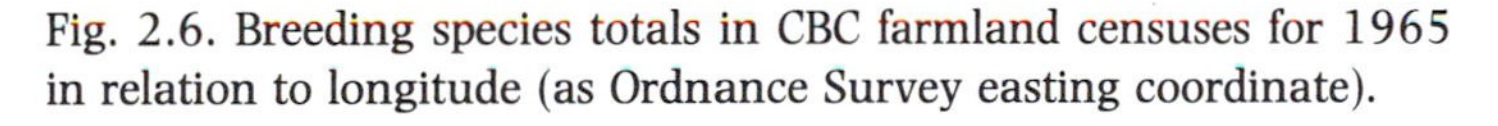

Fig. 2.6. Breeding species totals in CBC farmland censuses for 1965 in relation to longitude (as Ordnance Survey easting coordinate).

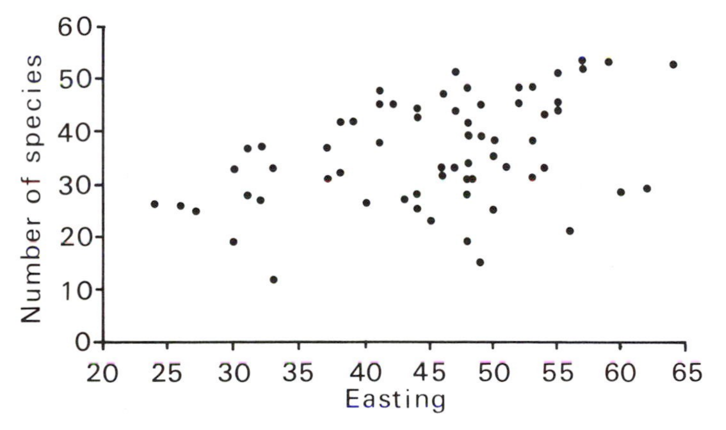

whilst numerical abundance (summed over all species) decreases from south to north (Fig. 2.7), in turn without any correlated change in species richness. Although altitude is generally greater in the west and north of Britain than elsewhere, neither trend is statistically correlated with altitude. Throughout Britain, therefore, both in the north and in the south, farms in the east are on average richer in species than are farms in the west. However, southern Britain is wider than is Scotland, and southern farms

Fig. 2.7. Geographical variation in bird density on British farmland. Data are averages for all CBC plots within the 100 × 100-km squares of the Ordnance Survey grid in 1965.

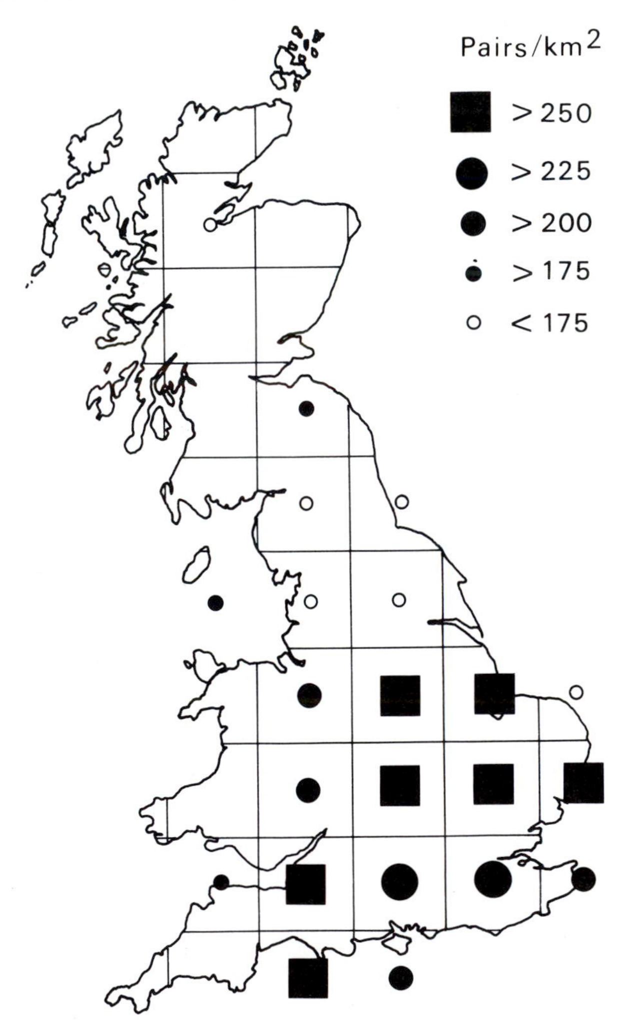

are more variable in the numbers of species breeding than are Scottish farms.

Williamson (1967) analysed Common Birds Census data on a regional basis, in an attempt to identify characteristics of bird communities nationally. We have extended his analysis to the whole of the CBC data for the years 1962–81, excluding plots with large edge effects (Verner, 1981; Marchant, 1981). We adopted a regional analysis based on the eight MAFF regions of England and Wales (Appendix 3), to facilitate the analysis of our results in relation to agricultural practice. For Scotland we classified counties into one of three (southern, central, and northern) regions, though the modest samples of Scottish CBC plots might well have been left amalgamated. We therefore considered a total of 11 regional samples.

Fuller (1982) suggests that a species may reasonably be regarded as 'typical' of a community if it is recorded in 50 per cent of the sample censuses of that community. On this basis, some 47 species covered by the CBC were typical of at least one of these 11 regions. Thirty-seven of these species were typical of the southern parts of Britain, and the remaining ten species were regular only on census plots in the north of England and in Scotland. Table 2.1 summarises the regional variations in the density of the most abundant species. The 12 species in this table were chosen on the basis of their regional averages exceeding 10 pairs per km^2 in at least one region. In most cases this also corresponded to the species being present in 90 per cent or more of the censuses for that region, the exceptions being very locally abundant species such as Willow Warbler (e.g. in the Yorkshire–Lancashire region) and Starling (e.g. in southern Scotland). The Blackbird was the only species to exceed this threshold density in all 11 regions, but four other species – Wren, Dunnock, Robin and Blue Tit – exceeded it everywhere except in part or all of Scotland, and the Chaffinch fell below this density only in the Yorkshire–Lancashire region.

Fig. 2.8 maps these regional averages and reveals a number of interesting geographical patterns. Four species – Blackbird, Song Thrush, Blue Tit and Great Tit – display simple south-north declines in abundance, already remarked as true of total bird densities on CBC plots in Britain. Wrens also become scarcer the further north the farm lies, but superimposed on this is a clear east-west increase in average density, with the south-west holding the greatest numbers. This pattern – previously remarked by Williamson (1967) – probably relates partly to agricultural factors and partly to the severity of winter conditions. On the first point, hedgerows are less frequently a component of field boundaries in the north (Pollard *et al.*, 1974), so hedgerow species find fewer suitable habitats there. On the

Table 2.1. *Regional variation in the densities of various species censused on farmland in Britain, averaged over 1962–81*

	Mean density (pairs per km²)										
	East	South-east	East Midlands	West Midlands	South-west	North	Yorkshire/ Lancashire	Wales	Southern Scotland	Central Scotland	Northern Scotland
Skylark	24.4	18.2	23.0	17.8	8.0	19.3	22.6	7.9	8.0	10.3	12.9
Wren	17.0	22.7	13.7	19.5	32.3	12.2	12.2	22.1	9.5	7.0	7.2
Dunnock	22.5	21.9	19.4	23.6	18.1	11.9	16.0	17.8	10.8	12.8	8.3
Robin	16.1	20.8	13.4	22.3	28.3	12.3	14.6	32.6	11.1	6.8	7.3
Blackbird	34.2	42.7	29.2	41.7	38.6	26.1	26.6	34.4	23.7	21.5	11.3
Song Thrush	16.0	17.4	10.7	13.8	9.1	11.7	10.6	13.5	11.0	8.2	6.4
Willow Warbler	4.7	5.2	8.9	8.9	8.9	18.7	13.6	19.2	13.8	16.3	8.8
Blue Tit	13.4	18.8	12.0	14.8	19.3	11.7	11.0	12.6	8.5	5.5	2.8
Great Tit	7.4	11.8	5.2	7.7	10.2	7.1	3.7	8.7	2.4	1.8	2.0
Starling	8.0	5.5	6.8	4.8	6.5	10.2	5.6	8.5	10.3	5.0	3.6
Chaffinch	14.6	21.7	15.8	19.5	31.0	28.9	9.8	35.1	29.9	21.4	20.4
Yellowhammer	12.5	8.0	14.5	13.8	6.8	9.4	5.0	7.5	5.0	9.6	12.3

Densities were derived as estimated on CBC census plots and are therefore subject to some error (Appendix 1).

Fig. 2.8. Regional variation in the densities (pairs/km²) of various species in 1965. From left to right and top to bottom: Skylark, Wren, Dunnock, Robin, Blackbird, Song Thrush, Willow Warbler, Blue Tit, Great Tit, Starling, Chaffinch, and Yellowhammer.

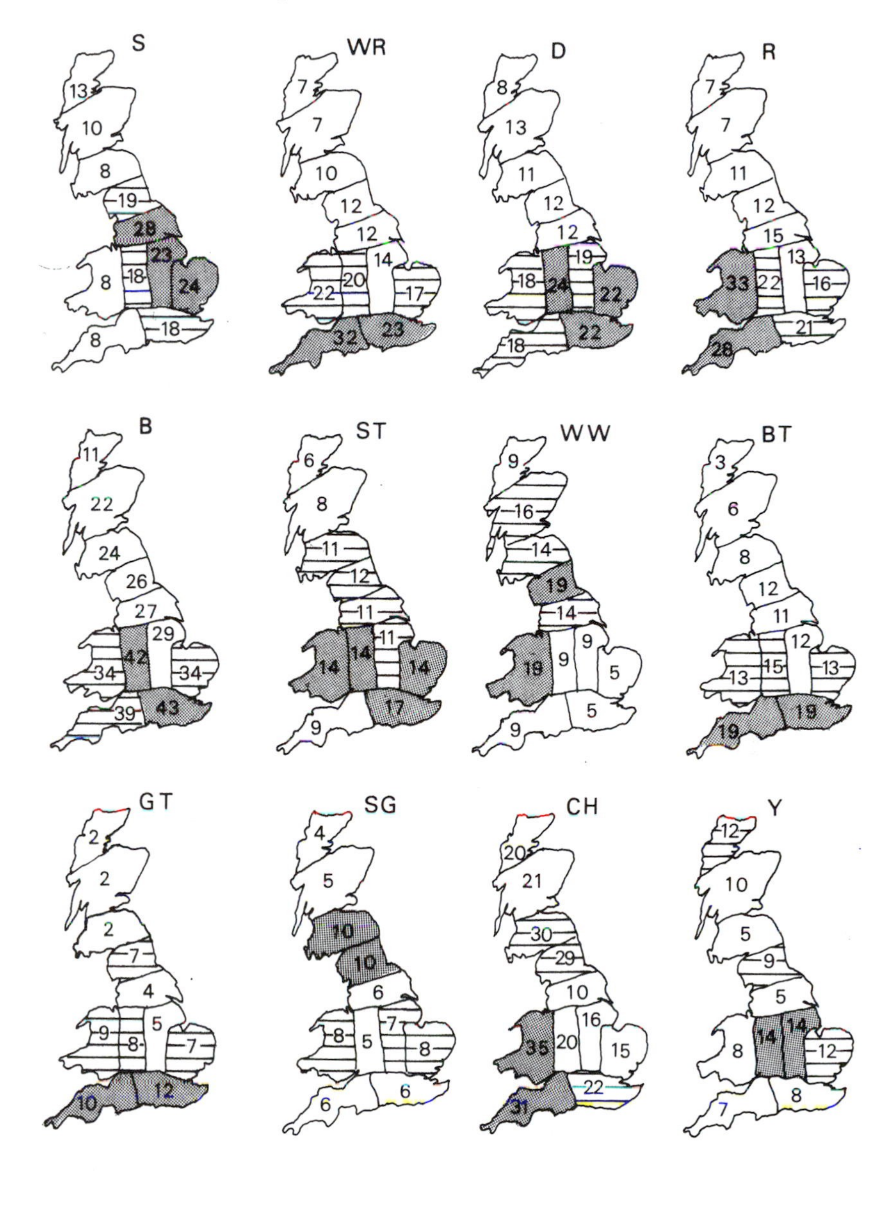

second point, Cawthorne & Marchant (1980) have shown that overwinter survival during the severe winter of 1978–79 was greatest in the mild south-west. The rather similar pattern for Robins – differing from Wren only in the relatively higher densities on the Welsh and West Midlands plots – may reflect the same considerations, for the Winter Atlas (Lack, 1986) also shows a winter concentration into these southern areas. An east-west increase in breeding densities is also apparent for Chaffinch and Willow Warbler but for these species the increase is probably associated with the greater availability of breeding habitat in the pastoral west, for Skylarks show the opposite trend, favouring the arable east (cf. Williamson, 1967). Within this region, leys are more often short-term and therefore particularly attractive to Skylarks (see Chapter 8). Regional variations in cropping practice probably also influence the distribution of Chaffinch, Skylark and Yellowhammer in the northern half of the country. Finally, outside the Scottish Highlands the higher densities of Starlings tend to be where the area of pasture is high or where stocking rates are low. In summary, therefore, the pattern of breeding densities for the more abundant species is largely explicable in terms of climate and agricultural practice.

Table 2.2 summarises the extent to which the most numerous species is different in different parts of the country. The vast majority of all farmland communities have Blackbird, Skylark or Chaffinch as their most abundant CBC species (i.e. omitting consideration of Woodpigeon, Rooks, Starlings and House Sparrows which are under-represented in the CBC (see Appendix 1)). In Wales, however, a fifth of all census lists were headed by the Robin. Wren and, surprisingly, Moorhen were the other dominants here, the latter reflecting locally good habitat. Blackbirds are usually the most abundant species on farms in the southern half of the country, but are slightly less so in Wales and the south-west, where Chaffinches are more often the more numerous species. In Scotland the Chaffinch occupies top place in half or more of the plots. The Skylark reached its greatest dominance, not surprisingly, in the arable east England region.

The less numerous species also vary regionally. Using the 'present in 50 per cent or more' criterion of Fuller (1982), some 17 species would be regarded as typical in the northern Scottish region (the region poorest in species), 19 species would qualify in Wales, 23 in Yorkshire-Lancashire, and between 27 and 34 elsewhere. The richest region is in central Scotland where species such as Oystercatcher, Curlew, Meadow Pipit and Redpoll are present on a majority of the plots to supplement the more southern species that also occur (though in smaller numbers the further north the

Table 2.2. *Regional variation in the extent of numerical dominance of CBC plots by various species*

	Eastern England		South-east		East Midlands		West Midlands		South-west		Northern England		Yorkshire/ Lancashire		Wales		Southern Scotland		Central Scotland		Northern Scotland	
	N[a]	%[b]	N[a]	%[b]	N[a]	%[b]	N[a]	%[b]	N[a]	%[b]	N[a]	%[b]	N[a]	%[b]	N[a]	%[b]	N[a]	%[b]	N[a]	%[b]	N[a]	%[b]
Blackbird	179	52.5	259	56.6	93	48.7	136	65.7	125	42.5	21	22.6	58	28.3	27	38.0	14	20.3	7	29.2	0	0.0
Skylark	81	23.8	58	12.7	52	27.2	22	10.6	10	3.4	21	22.6	85	41.5	(2)	2.8	7	10.1	2	8.3	7	30.4
Chaffinch	3	0.9	50	10.9	10	5.2	18	8.7	69	23.5	34	36.6	13	6.3	15	21.1	37	53.6	8	33.3	12	52.2
Wren	22	6.4	21	4.6	—	—	12	5.8	48	16.3	—	—	—	—	4	5.6	—	—	—	—	—	—
Robin	—	—	—	—	—	—	—	—	—	—	—	—	12	5.8	14	19.7	—	—	—	—	—	—
Dunnock	20	5.9	—	—	—	—	—	—	—	—	—	—	—	—	—	—	—	—	—	—	—	—
Yellowhammer	—	—	—	—	12	6.3	—	—	—	—	—	—	—	—	—	—	—	—	—	—	3	13.0
Moorhen	—	—	—	—	—	—	—	—	—	—	—	—	—	—	5	7.0	—	—	3	12.5	—	—
Sum of first three species	—	77.2	—	80.2	—	81.1	—	85.0	—	69.4	—	81.8	—	76.1	—	61.9	—	84.0	—	70.8	—	82.6

[a] Number of censuses in which this species was the most numerous present.
[b] Per cent of standard CBC censuses headed by this species.

plot lies). The next richest region is in eastern England. Extensive areas of East Anglian farmland have been reclaimed from marshland or from heathland, and fragments and relicts of these habitats and of woodland must contribute to the greater abundance of bird species on these easterly farms. For example, Williamson found that the presence of a network of drainage ditches on low-lying fenland farms resulted in high numbers of some aquatic species: Mallard and Sedge Warbler were recorded in particularly exceptional densities.

Williamson (1967) discussed the regional variations in farmland bird communities in some detail, using a sample of Midlands farms to provide a standard for comparison. Most of the more marked variations he noted relate, as might be expected, to the less numerous species. In his 'Eastern Counties' group (Lincolnshire, Cambridgeshire, Norfolk, Suffolk and Essex), the striking features were high densities of gamebirds and of Turtle and Stock Doves. Gamebirds are particularly numerous in the east partly because of the scale of game preservation and partly because of the abundance of Red-legged Partridges, encouraged by the more continental climate of the region. Game preserving probably also accounted for the scarcity of Magpies noted by Williamson, since this species has since increased with the decline in gamekeepering there (Tapper, 1981; see p. 184).

For the south and south-west of England Williamson combined the data from CBC plots from the Channel counties from Kent to Cornwall, together with Surrey, Wiltshire and Somerset. Here the interesting feature was the importance of fragments of other habitats rather than the presence of geographical variation *per se*. This was a particular advantage to wetland birds, and farms with areas of marsh or reedbed had good populations of Moorhens, Coots, Reed and Sedge Warblers, and Yellow Wagtails. Such farms also held high populations of Wrens and Song Thrushes.

In the West Midlands and Wales (Herefordshire, Gloucestershire, Glamorgan, Radnorshire, and Monmouthshire) Williamson found that the greater abundance of hedgerow trees there led to high densities of Robins, Whitethroats, Dunnocks and Chaffinches. Densities were otherwise similar to those of the Midlands but some species were more widespread in the west. These included Stock Dove (present on 40 per cent of farmland plots), Little Owl (60 per cent), Long-tailed Tit (53 per cent), Wren (100 per cent), Redstart (53 per cent) and Meadow Pipit (26 per cent).

Farmland in northern England is ecologically very different from the south, and field and steppe species were correspondingly more prominent than in the Midlands (Williamson, 1967). Curlew and Snipe occurred in

75 per cent and 52 per cent of the censuses respectively and Lapwing and Meadow Pipit were both regular, the Lapwing being especially numerous on farms in the Pennines. Five other wader species bred and breeding wildfowl included Teal and Goosander as well as the more widespread Mallard. Williamson found that most hedgerow species were relatively scarce, however, this being balanced by a greater variety of breeding passerines. Pennine farms were particularly rich in passerine species, with Wheatears, Whinchat, Redpoll and Redstart each relatively numerous there.

In general our own longer term analyses support Williamson's conclusions but there have been important changes since he wrote. Thus the regional patterns of Stock Doves noted by Williamson (1967) proved to be transitional (O'Connor & Mead, 1984). The species was badly affected by the widespread use of dieldrin and other organochlorine seed-dressings in the 1950s. Egg success and adult survival were particularly affected in arable areas and the population contracted towards the coast (Chapter 9). After some of the chemicals concerned were withdrawn in 1962 a recovery began, relatively faster in southern Britain than in eastern Britain. In the west, however, Stock Doves are now more local and it is possible that the loss of hedgerow trees to Dutch elm disease may be implicated (p. 144). The recorded frequency of Little Owls, another species vulnerable to loss of hedgerow trees, has also fallen since Williamson wrote.

In the north the Skylark is no longer the commonest species. Our analyses suggest that its dominance in Williamson's surveys there was because other species were more severely affected by the hard winter of 1962–63. In the longer term the Chaffinch has emerged as the most numerous species on northern farms, followed by Blackbird, then Skylark. Interestingly, we found that Blackbirds more frequently dominate bird-rich farms there whilst Skylarks more frequently dominate bird-poor farms. Following the general recovery of most species since the 1962–63 winter, the 'core' species of northern farmland are now much the same as in other regions, and only the Meadow Pipit is unusually abundant there.

Influence of agricultural practice

Farmland in Britain ranges from intensively managed specialist cereal enterprises through to Scottish and Welsh hill farms, and their bird communities are correspondingly varied. Against this background Williamson classified farms as mainly arable (if more than 60 per cent of their cultivated areas were under cereals or root crops) or as mainly pasture (otherwise). Most hedgerow species had similar densities in both types of farm but the densities of partridges, Skylark and Reed Bunting were two to three times greater on the arable farms. We examined the CBC data on a somewhat different basis, because of our interest in the effects of agricultural practice on bird populations. We have considered four samples of CBC plots, drawn from counties we classified as dominated respectively by tillage, by cereal production, by sheep rearing, and by pasture management (Appendix 3).

Fig. 2.9 shows the frequency distribution for species richness in relation to primary agricultural practices. These four practices are those that can be most distinctively defined from the available agricultural statistics. In each case the number of species recorded in censuses varied from 10 or

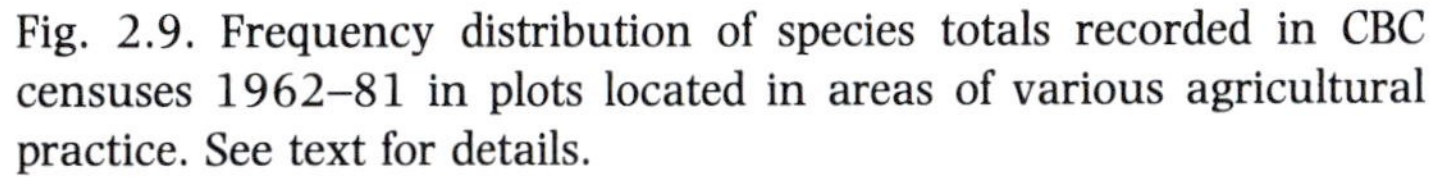

Fig. 2.9. Frequency distribution of species totals recorded in CBC censuses 1962–81 in plots located in areas of various agricultural practice. See text for details.

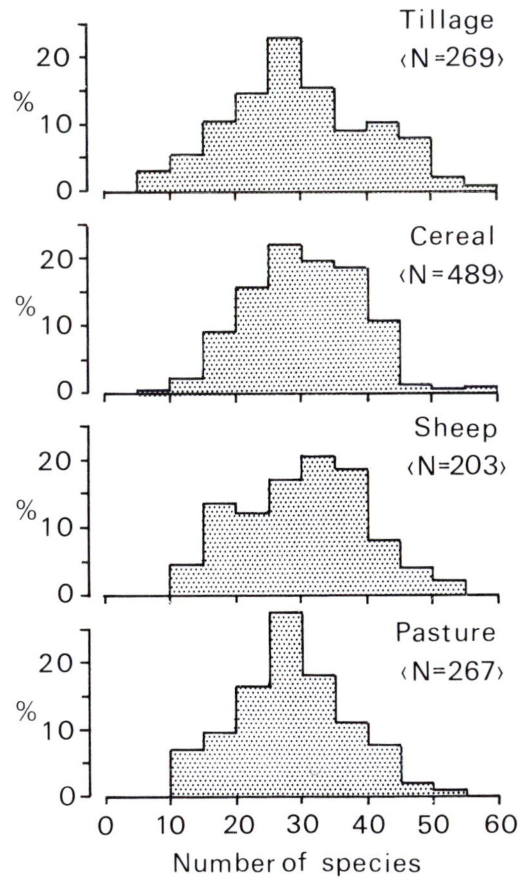

fewer up to a maximum of between 50 and 60 species. These counts relate only to the standard CBC census species. Censuses in the sheep-rearing areas typically recorded 30–35 species, and the other areas 25–30 species, but within all four groups individual farms varied greatly as to how many species were present.

The total density of birds on the various plots has varied markedly over time, primarily due to the severe winter of 1962–63 but smaller falls in population levels occurred in 1978–79 and 1981–82, the other two severe winters since 1962. On tillage plots, for example, total density has gone from a minimum of 140 pairs per km^2 in 1963 to just under 340 pairs per km^2 in 1976. Against this background the differences in mean densities between the four samples are small, though there is a fairly consistent tendency within any single year for sheep-rearing areas to hold the highest densities, with cereal plots in second place and either tillage or pasture plots holding the lowest densities.

Eight species – Skylark, Wren, Dunnock, Robin, Blackbird, Song Thrush, Blue Tit and Chaffinch – were among the 10 most abundant species in all four environments. Three other species – Linnet, Yellowhammer and Reed Bunting – ranked in the top ten (the last tying tenth within tillage) in the two arable environments. The Willow Warbler was numerous in the two pastoral environments, and Great Tit and Starling were in the top ten only in sheep-rearing and pasture areas respectively. The Starling may be under-represented in these CBC samples because of its locally clumped breeding distribution. Although the same few species are the most abundant in all four environments, their relative rankings differ between the four groups, particularly between arable and pastoral areas. Blackbird, Robin, Wren and Chaffinch occupy the top four places in the two pastoral environments but in arable areas Skylark and Dunnock are second only to Blackbird. The Dunnock is rather tolerant in its habitat requirements, a feature which has probably enabled it to adapt to arable land; its densities there are about one-third higher than in pastoral areas. The Skylark is about twice as abundant on arable land as on sheep or pasture lands. On tillage it forms about 14 per cent of the community but elsewhere it accounts for just 5–8 per cent.

Table 2.3 examines whether dominance varied between the four agricultural regimes. Three species – Blackbird, Skylark and Chaffinch – were usually the most abundant species found within individual censuses. Some two-thirds of the cereal communities were headed by Blackbirds, and Skylark topped about half the remainder; the balance was shared between Chaffinch, Wren, Dunnock, and a handful of other species. Plots in tillage counties were as likely to be headed by Skylark as by Blackbird, with Dunnock ousting the Chaffinch from third place. About a third of the pasture management plots were topped by Blackbird, and Chaffinch and Skylark were almost equally represented. The remaining one-third of the plots were headed by any of several species, including Wren, Robin, Blue Tit, Dunnock and Moorhen, a diversity reflecting the variety of habitat fragments to be found in pasture areas (Chapter 6). Finally, in sheep-rearing areas the Chaffinch was most frequently the dominant species, with Blackbird in second place and Wren, Robin and Skylark accounting for the bulk of the remainder. One reason for Chaffinch taking first place from Blackbird may lie in the generally higher altitudes of the sheep-rearing areas: Blackbirds are scarce above about 500 feet (152 m).

Table 2.4 looks at the communities headed by Blackbird, Skylark and Chaffinch respectively, to see what share of these communities each of them held. In the cereal censuses topped by Blackbirds (two-thirds of them) the species accounted for, on average, 14.4 per cent, i.e. one bird in seven on these plots was a Blackbird. Where either Skylark or Chaffinch was the most numerous species on a cereal plot, however, they were only as frequent as one in 12 or one in 13 birds on the plot. In the pasture plots the top species generally constituted from one in 10 to one in 13 birds in the community, whilst in the sheep-rearing areas the plots dominated by Chaffinch or Blackbird had one in nine or one in 10 birds as the top species. Where Skylarks topped the list they accounted for only one in 20 of the birds on the plot concerned.

Table 2.3. *The frequency of numerical dominance by different species on CBC plots in areas of different agricultural practice*

	Plots (percentage) on which species is dominant in areas of							
	Tillage		Cereals		Sheep		Pasture management	
	N	%	N	%	N	%	N	%
Blackbird	112	41.3	319	65.2	49	24.1	89	33.3
Skylark	101	37.3	81	16.6	19	9.4	43	16.1
Chaffinch	4	1.5	30	6.1	69	34.0	47	17.6
Dunnock	16	5.9	10	2.0	1	0.5	8	3.0
Wren			17	3.5	22	10.8	24	9.0
Robin					16	7.9	12	4.5
Blue Tit							12	4.5
Moorhen							13	4.9

As one might expect from these figures, in most communities headed by one of Blackbird, Skylark or Chaffinch one of the other two species ranked second. However, Dunnock, Robin and Wren, although rarely the most numerous species on a plot, were frequently in second place, more or less regardless of the main agricultural practice. Unusually, Moorhens also appeared in second place, in about 10 per cent of those tillage censuses dominated by Blackbird and in about a quarter of the sheep plots dominated by Chaffinch or Skylark. The Moorhen is one of the species whose abundance is markedly dependent on the availability of small ponds and waterways (Chapter 6) and can achieve high numbers where these are present. Skylark-dominated plots were generally the most diverse in species composition, particularly under pasture (where Meadow Pipit or Lapwing were the second most numerous species in about half the censuses) and under tillage (where some 29 per cent of the plots had one or other of the three common buntings as the second most abundant species).

Our results thus confirm the conclusions of Wyllie (1976) who suggested that most farmland supports about six dominant species and 19 common or sub-dominant species, these between them accounting for half of all species breeding on the farm. Our results show that cereal and tillage areas are slightly richer in number of species than are areas under sheep or pasture, and the cereal areas possibly sustain more individuals than do other regimes. Tillage plots are clearly rather varied in the composition of the bird communities present. Sheep-rearing areas support a wide range of species, some very numerous but others rather scarce, perhaps largely on account of the altitude of many of the sheep counties. The tillage and cereal plots are typically strongly dominated by Blackbird or Skylark whilst

Table 2.4. *Mean relative abundance of Blackbird, Skylark and Chaffinch in those CBC censuses in which they are the most abundant species*

	Percentage of community in areas of							
	Tillage		Cereals		Sheep		Pasture management	
	Mean	s.e.	Mean	s.e.	Mean	s.e.	Mean	s.e.
Blackbird	11.2	0.29	14.4	0.22	10.2	0.31	10.5	0.30
Skylark	13.7	0.70	8.5	0.36	5.4	0.39	8.0	0.61
Chaffinch	2.7	0.14	7.0	0.19	11.4	0.33	8.3	0.35

pastoral areas hold communities of greater variety, both within and between plots.

Farmland birds in winter

Information on winter bird communities is scarcer than for breeding communities. In winter both the resources available on farmland and the birds present there change. Murton (1971) drew attention to the seasonal increase in invertebrates which are largely unavailable during the dry period of mid-summer and to the consequences of this in supporting an influx of winter immigrants of Lapwings, Starlings, thrushes and other ground-feeding species (p. 15). For grazing and seed-eating species, too, autumn harvesting may leave substantial crop residues such as stubble grain and waste potatoes whilst autumn sowing creates young growth of corn or grass that may be grazed in late winter by geese. Spencer (1983) has recently reviewed the ways in which birds may adapt to the food sources available during winter.

A striking feature of the winter bird community on farmland is the difference from the breeding community in the species complement present on each farm. In addition, individual species are markedly clumped into foraging flocks in winter. Such flocking is largely an adaptation to difficulties in locating the clumped food sources of winter. But the change in the species present is a more complex phenomenon, due in part to the departure of summer and partial migrants but due even more to upland species such as Golden Plover and Meadow Pipits descending to the lowlands to winter and to the influx of surface-feeding winter visitors from northern Europe already mentioned. Given such diverse origins, it is clear that the richest bird communities in winter should be those offering the greatest diversity of habitat, so that mixed farming should offer a wider range of resources than more homogeneous areas. That this is indeed the case is suggested by a comparison of preliminary results from the BTO/IWC Atlas of Winter Bird Distribution and the geographical distribution of mixed farming in Great Britain (Fig. 2.10).

In Fig. 2.10 the Winter Atlas data show the geographical distribution of species richness in winter, based on the top (most abundant) 20 per cent of squares for each species; the high spots on the map thus indicate areas where most species are most abundant in winter. The index of mixed farming presented in the same figure plots the ratio of land under tillage and under grass in each county, without regard to which use is dominant. Thus, a county with a 35:65 ratio of tillage to grass has the same index as one with a 65:35 ratio. Comparison of the two maps (Fig. 2.10) shows

Fig. 2.10. The distribution of wintering species in Britain (left) and the grass: tillage balance in different regions (right). The wintering bird data are for the Winter Atlas and show the number of species recorded in any given 10-km square. The grass: tillage ratio is expressed in categories irrespective of whether grass or tillage was dominant, i.e. it is the balance or imbalance that is described.

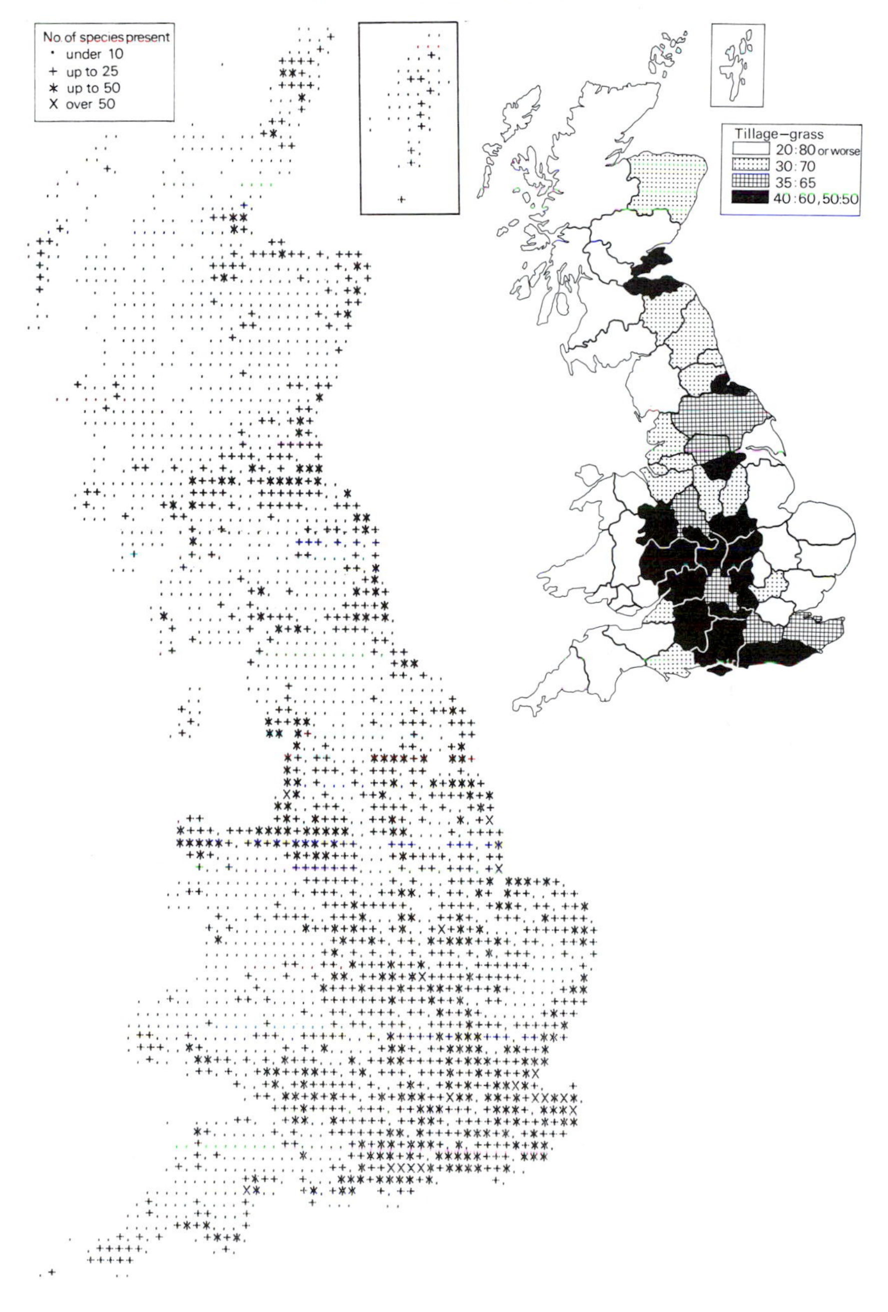

a strong correlation between the two, including concentrations of winter birds in local areas of mixed farming in South Yorkshire, around Newcastle, and around the Firth of Forth. The Winter Atlas map shows some coastal concentration of species not matched by the agricultural map, but this is the result of including all species, including ducks, shorebirds and seabirds, in our analysis. A few apparent mismatches are due to variations in agricultural practice within individual counties, because the Winter Atlas data are on a finer scale than the county-wide agricultural statistics. For example, farmland in east England is shown to be much more favoured than the south-west and Wales, where farming is dominated by grassland. The main characteristic of the farming of the eastern counties, however, is the wide range of crops and, as a consequence, the different types of land management practised. This variation within the primary classification of arable land is another valuable form of diversity for birds which allows them to exploit the more varied methods and seasonal patterns of work and crop residues. A similar relationship between variety of cropping and the distribution of breeding birds has been observed on arable land in Switzerland (Luder, 1983).

The separate distributions of stock, tillage, and other crops examined do not show the strong correlation with winter bird distribution shown in Fig. 2.10. We conclude, therefore, that a mixture of both tillage and stock farming is better for birds in winter than is either alone. This result generalises the conclusion of Brenchley (1984), who found that the distribution of Rooks in Great Britain has historically correlated with areas of mixed agricultural practice, and that a ratio of 45 per cent tillage and 55 per cent grass was optimal. For the resident Rook this correlation is largely determined by the availability of different food sources at different times of year (Macdonald, 1983; Brenchley, 1984) but here the correlation is more likely to be because mixed farming provides more species with suitable resources through the winter. The extent to which populations of the resident species are limited by the resources found on farmland in winter is at present unknown.

Summary

Farmland bird communities consist of a small number of very abundant and widespread 'core' species and about 20 less abundant but fairly widespread species; other, rarer species are incidentally present in odd patches of suitable habitat. Farming practice varies regionally due to geographical variation in soil, climate and relief. These factors may themselves limit the distribution of some species but regional variation in farmland bird communities is also affected by regional agricultural practice. Blackbird, Chaffinch and Skylark are the most numerous farmland species, though Rook and Woodpigeon are particularly important as a proportion of community biomass. Winter bird communities have been poorly studied but in many species the highest densities are found where mixed farming is at its most diverse.

3: Farmland as a nesting habitat

Rather little attention has been given to the value of farmland as a nesting habitat, though some species studies contain much relevant information (e.g. Murton, 1958; Newton, 1967; Potts, 1980). An understanding of the current ecology of farmland breeding birds is much needed, for at least three reasons. First, it has been argued that nesting in hedgerows by many small passerines is very unproductive and that only individuals able to nest in woodland habitats produce many recruits to future generations (Murton & Westwood, 1974). In this context the work of Snow & Mayer-Gross (1967) is the major review of farmland as a breeding habitat. Second, the widespread use of organochlorines in the 1950s markedly reduced the breeding success, especially hatching success, of pairs that escaped direct poisoning (e.g. Newton, 1974), so recent trends are of interest. Finally, recent changes in the management of farm crops have implications for breeding success on agricultural land (Chapter 8). We cannot attempt here a full review of nesting ecology in a farmland context but we provide a brief overview of relevant aspects of nesting biology as background to more detailed discussion in later chapters.

Much of the information we present has been obtained through fresh analyses of data in the BTO Nest Records Scheme. This is described in greater detail in Appendix 2, where an account of our methods is presented, but two important points need to be made here. First, the nest record cards provide no information on post-fledging survival, leaving an important gap in our knowledge of habitat-specific success. Second, many previous studies have assessed breeding success as the proportion of nests from which at least one young fledged (e.g. Snow & Mayer-Gross, 1967). This definition gives an optimistic estimate of how good a habitat is for birds, particularly where nestling starvation is important. Suppose two habitats each suffered, say, 25 per cent nest predation but in one the parents at the surviving nests can successfully feed four chicks whilst in the other they can feed only a single chick. The standard, nest-based definition of success says that the two habitats are equally good: each achieves 75 per cent success. Yet the first produces four times as many young as does the latter. We therefore computed hatching success (chicks hatched per egg laid) and fledging success (chicks fledged per hatchling) and multiplied these values to estimate breeding success (chicks fledged per egg laid). Where predation is the major cause of nest failure, as in the study by Snow & Mayer-Gross (1967), the two definitions give very similar results but

Table 3.1. *Clutch size in various habitats, 1962–80*

	Clutch size					
	Farmland		Suburbia		Woodland	
	Mean	s.d.	Mean	s.d.	Mean	s.d.
Dunnock	3.8	0.82	3.9	0.89	3.9	0.77
Blackbird	3.8	0.82	3.7	0.86	3.8	0.83
Song Thrush	4.2	0.72	4.1	0.70	4.1	0.68
Mistle Thrush	3.9	0.79	3.9	0.62	3.9	0.72
Whitethroat	4.7	0.72	4.6	0.87	—	—
Magpie	5.7	1.30	5.3	1.53	5.3	1.49
Jackdaw	4.3	1.05	4.3	1.05	—	—
Rook	4.0	1.10	—	—	—	—
Starling	4.6	0.91	4.4	0.78	4.6	1.06
Chaffinch	4.3	0.78	4.2	0.84	4.2	0.81
Greenfinch	4.7	0.76	4.7	0.87	4.7	0.85
Linnet	4.8	0.66	4.6	0.72	4.7	0.73

Dashes indicate small samples.
Data are from BTO nest record cards.

when feeding conditions vary substantially between habitats, our procedure is more satisfactory.

Table 3.1 examines clutch size in a number of species commonly found in farmland and in woodland or suburban habitats. Clutch size does not vary significantly between habitats in any single species but in 10 of the 12 species listed clutches were larger on farmland than elsewhere. The exceptions were Greenfinch, in which clutches were very similar in all three habitats, and Dunnock, in which farmland clutches averaged about 3 per cent smaller than elsewhere. There is thus a consistent trend for farmland clutches to be slightly larger than those in suburban and urban areas.

These minor trends in respect of clutch size are dwarfed by habitat differences in egg and nestling mortalities (Table 3.2). The thrushes showed the most marked habitat differences: for Blackbird, mortality of both eggs and chicks was greatest in suburban areas; Song Thrushes, on the other hand, did best there, particularly for eggs. Mistle Thrush suffered high mortality on farmland only in eggs; for chicks, suburban areas were the worst habitat. Both the Rook and the Jackdaw did better in woodland than on farmland. Linnet and Greenfinch suffered more in suburbia than they did on farmland. Starling and Dunnock did less well in woodland. Finally, mortalities were lower in Whitethroat nests on farmland than anywhere else.

These differences are not as clear cut as were those detected by Snow & Mayer-Gross (1967) who found, for example, that breeding success of Blackbirds and Song Thrushes on farmland was intermediate between the high success in suburban gardens and very low success in woodland, differences they thought were probably associated with the density of nest predators in the various habitats. In these species, nesting success increases in all habitats as the season progresses, probably in part because of improving vegetation cover and in part because nest predators have alternative foods available later in the season. Corvids, for instance, concentrate from mid-summer onwards on grain and weed seeds rather than on nestlings (Holyoak, 1968) and this results in a sharp seasonal reduction in predation. By way of example, one Sutherland study found that none of the Collared Dove clutches laid before mid-June escaped predation by Hooded Crows but two-thirds of the clutches laid after this date were successful (Macdonald, 1977).

Snow & Mayer-Gross (1967) found that the breeding success of the Greenfinch was fairly similar in all major habitats, nest success averaging 36–38 per cent in farmland samples, 37 per cent in all rural habitats, and 35 per cent in all suburban/urban habitats. These results are in contrast

Table 3.2. *Egg and chick mortality in various habitats, 1962–80*

	Egg mortality, % per day			Chick mortality, % per day		
	Farmland	Suburbia	Woodland	Farmland	Suburbia	Woodland
Dunnock	4.4	4.6	4.5	6.0	5.7	6.7
Blackbird	4.0	4.6	3.7	5.0	6.0	5.5
Song Thrush	4.1	3.6	4.1	6.2	5.6	6.3
Mistle Thrush	4.6	3.8	4.0	5.3	5.5	5.0
Whitethroat	2.9	3.6	4.8	6.0	6.9	6.2
Magpie	4.1	3.9	4.6	2.3	2.1	2.2
Jackdaw	2.8	2.7	—	2.8	3.1	2.4
Rook	4.0	—	1.9	1.6	—	1.2
Starling	3.1	3.1	3.3	3.2	3.9	4.1
Chaffinch	4.2	4.3	4.0	6.2	5.9	7.0
Greenfinch	4.1	4.8	4.3	4.5	5.3	5.5
Linnet	3.8	4.5	3.8	5.6	6.8	6.4

Dashes indicate small samples.
Data are from BTO nest record cards.

to our data, which show that the Greenfinch does best nowadays on farmland. Snow & Mayer-Gross also noted a large difference in Chaffinch nest success on farmland and in woodland (17 per cent versus 11 per cent) which they thought was due to the conspicuousness of the nest and the high frequency of predators in woodland. However, our data show a smaller difference (24 per cent versus 22 per cent) when success is estimated in terms of fledglings per egg laid. This is due in large part to a reduction in egg predation over the last decade (see below).

Clutch size varies seasonally on farmland, partly depending on the availability of food. Blackbirds nesting on farmland make considerable use of earthworms, which become less available as the season advances. This is in contrast to the situation for Blackbirds in woodland where a variety of caterpillars and other invertebrates become numerous in early summer, permitting a seasonal increase in clutch size in that habitat. During a drought in May 1961, 8 out of 17 broods of Blackbirds died in the nest in farmland near Oxford but some young were reared from all 10 woodland nests being studied at the same time. For other species Snow & Mayer-Gross (1967) found clutch sizes to be relatively uniform over different habitats.

In many other farmland species clutch size decreases steadily through the season, as with the Whitethroat (Fig. 3.1), again most likely owing to changes in food availability. In Whitethroat, for example, partial losses of clutches increase as clutch size decreases through the season, suggesting

Fig. 3.1. Mean clutch size (± SE) of Whitethroats throughout the breeding season. After Mason (1976).

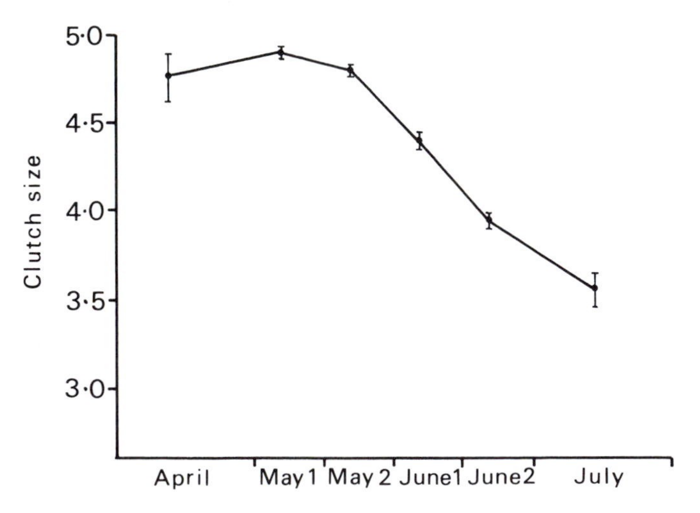

that even the reduced clutches are slightly larger than can be fed properly (Mason, 1976). Nest success increases over the same period, however, as vegetation develops and conceals the nests from predators to a greater degree. Nest productivity is therefore greatest in mid-season (in June in the case of the Whitethroat). In other species, though, seasonal changes in clutch size are unlikely to be wholly due to food. In a Scottish study of Swallows, for example, clutch size decreased with laying date while first clutches were being laid but rose again at the start of second clutches (which once more declined with date); nevertheless, early second clutches were slightly smaller than early first clutches, and late clutches followed a similar pattern (McGinn & Clark, 1978). Here the temporal decline is more likely to be due to older, more experienced birds starting earlier and laying large clutches (both first and second), and to younger birds starting later and laying fewer eggs.

Fig. 3.2. Difference in incidence of predation of nests: (top) between farmland and woodland for 21 species; (bottom) between farmland and suburban areas for 16 species. Predation rates as percentage of nests lost to predators over the entire nest cycle. Nest record card data, 1962–80.

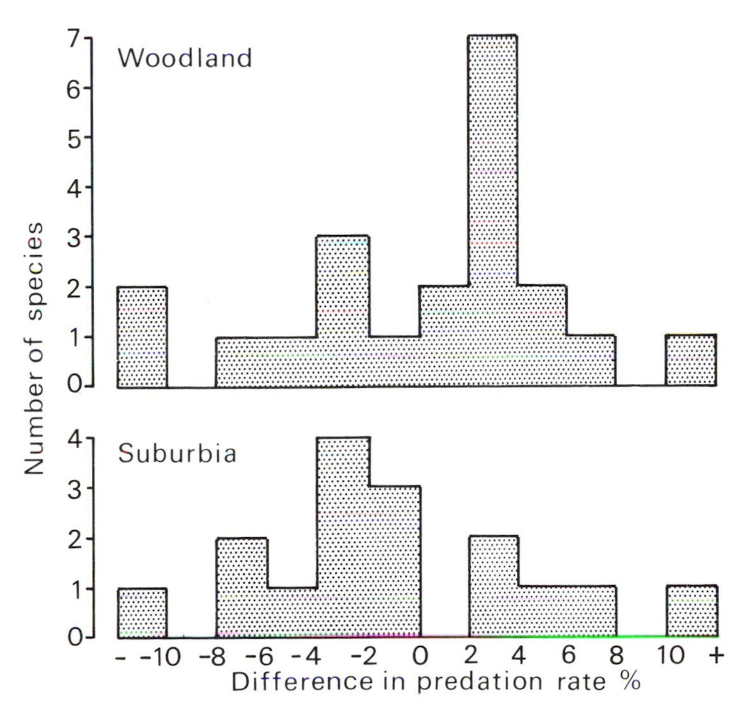

In many ways farmland with its mixture of linear and patchy cover and lower density of nest predators should be intermediate between woodland and suburban gardens, and nest success as measured by Snow & Mayer-Gross (1967) does seem to reflect this. We therefore examined the nest record cards for various species and computed the incidence of observed nest predation in the three habitats. Species were those of Table 3.2 except Jackdaw. We found first that predation frequency in different species was correlated across habitats, so that a species such as the Chaffinch experiences heavy predation in all three habitats and a species such as the Starling experiences only light predation in all three habitats. Secondly, we found that superimposed on these correlations across species was a between-habitat effect (Fig. 3.2). Predation is more rife in woodland than in farmland, and is more rife in farmland than in suburban areas. Thus farmland lies midway between woodland and suburbia in terms of predator impact. However, this is a general pattern, for there can be strong local differences within habitats: the hedges and gardens around farmhouses, for example, frequently hold concentrations of the breeding birds on a farm (p. 115) but may suffer heavily from predation by farm cats.

Regional variation

In a number of species breeding performance varies across the country. The extent of these regional differences is different between species and also varies as to whether clutch size, egg mortality or nestling mortality is affected. Blackbird and Skylark, for example, show little more than 10 per cent variation in mean clutch size between regions but for Mistle Thrush clutch size varies from 2.9 eggs in eastern England to 4.5 eggs in the south-west (Table 3.3). Regional differences in clutch size were particularly pronounced in Magpie, Jackdaw, Starling, Kestrel, Chaffinch and Whitethroat but little consistency was apparent between species as to the quality of each region for breeding success (Table 3.4).

The amount of grass mown in each region has varied markedly since the early 1960s, when the BTO nest records analyses begin, but throughout this period the relative ranking of the different regions in the extent of mowing has remained remarkably constant. From an average taken over the 19-year period of the nest records, clear relationships are apparent. Fledgling production in the Mistle Thrush increases sharply with the extent of mowing (Fig 3.3), and clutch size increases in parallel. The result is that significantly more fledglings are produced by a pair of Mistle Thrushes breeding in the south-west (greatest mowing) than by a pair breeding in east England (least mowing), and intermediate areas are correspondingly

Table 3.3. *Regional variation within England and Wales of clutch size in various species*

	South-east England	South-west England	Eastern England	East Midlands	West Midlands	Wales	Yorkshire/ Lancashire	Northern England
Lapwing	3.75	3.83	3.72	3.63	3.67	3.68	3.61	3.78
Skylark	3.42	—	3.35	3.42	3.30	—	3.30	3.47
Dunnock	3.64	4.00	3.83	4.03	3.86	4.00	3.91	4.00
Blackbird	3.67	3.87	3.82	3.89	3.88	—	4.04	3.89
Song Thrush	4.14	4.40	4.19	4.13	3.97	4.50	4.24	3.90
Mistle Thrush	3.92	4.54	2.88	3.58	3.90	4.20	3.94	3.89
Whitethroat	4.79	4.67	4.50	4.67	4.74	5.00	4.83	5.09
Magpie	5.92	5.50	5.76	5.44	5.30	6.16	5.93	5.25
Jackdaw	4.28	4.39	4.18	4.25	4.70	3.80	4.18	4.80
Starling	4.52	4.50	4.92	4.33	4.49	5.43	4.54	5.00
Chaffinch	4.37	4.40	3.80	4.52	4.12	4.28	4.36	4.24
Greenfinch	4.87	4.83	4.63	4.52	4.91	—	4.59	4.56
Goldfinch	5.00	5.22	4.92	5.08	—	4.83	—	—
Linnet	4.78	4.77	4.72	4.84	4.83	—	4.86	4.83

Dashes indicate small samples.
Data are from BTO nest record cards.

Table 3.4. *Regional variation within England and Wales in egg and nestling mortality (per cent per day) of various species*

	Egg mortality							
	South-east England	South-west England	Eastern England	East Midlands	West Midlands	Wales	Yorkshire/ Lancashire	Northern England
Lapwing	2.2	2.5	2.1	3.4	2.8	4.0	3.3	2.6
Skylark	4.8	6.8	7.9	7.1	6.9	2.0	6.2	5.2
Dunnock	4.7	4.1	5.1	4.0	3.8	1.6	4.1	4.5
Blackbird	4.5	4.8	3.8	3.5	4.4	5.3	3.5	3.2
Song Thrush	4.6	4.6	3.7	3.1	3.7	5.1	4.2	4.7
Mistle Thrush	4.7	4.6	5.4	6.3	4.1	3.1	5.1	3.1
Whitethroat	2.4	1.7	3.3	3.2	2.3	3.6	3.8	2.8
Magpie	3.3	3.8	3.6	4.0	2.4	6.8	5.6	3.3
Jackdaw	2.2	3.2	2.1	2.7	2.5	2.0	3.4	3.9
Starling	3.4	2.0	3.5	3.3	2.9	7.1	4.3	1.5
Chaffinch	4.4	4.7	3.8	4.9	3.8	4.2	3.9	3.8
Greenfinch	4.9	2.2	4.4	4.2	4.2	—	4.0	2.6
Goldfinch	3.9	4.8	3.4	3.8	5.7	3.3	—	6.2
Linnet	3.5	3.4	3.2	4.2	4.1	10.8	4.3	4.0

Dashes indicate small samples.
Data are from BTO nest record cards.

Table 3.4 cont.

	Nestling mortality							
	South-east England	South-west England	Eastern England	East Midlands	West Midlands	Wales	Yorkshire/ Lancashire	Northern England
Lapwing	3.0	—	6.7	1.9	1.5	—	4.6	3.5
Skylark	9.9	8.2	12.7	6.2	9.3	6.6	7.4	9.8
Dunnock	6.6	5.4	6.1	4.8	6.6	5.5	6.6	5.6
Blackbird	4.8	4.0	6.1	5.0	5.1	5.4	5.7	4.5
Song Thrush	6.9	6.1	5.4	7.9	5.9	5.6	5.7	5.6
Mistle Thrush	5.3	5.2	6.6	5.7	4.7	6.6	5.3	5.8
Whitethroat	4.3	4.5	6.1	7.0	6.6	4.9	7.9	8.9
Magpie	2.2	2.8	2.0	2.4	1.6	2.1	2.3	4.7
Jackdaw	2.6	2.8	3.2	3.6	2.3	1.9	5.0	1.7
Starling	3.2	2.8	2.6	3.5	2.7	2.2	4.4	4.0
Chaffinch	6.1	5.7	7.5	5.7	5.7	7.7	6.2	6.3
Greenfinch	6.0	3.6	3.2	4.6	4.0	—	5.2	3.0
Goldfinch	4.8	7.5	5.9	6.3	2.8	3.7	—	—
Linnet	5.4	5.1	6.2	5.0	5.9	10.0	7.0	4.4

intermediate in productivity. As Mistle Thrushes will move long distances from their nest to forage over grassland it is perhaps not surprising that such a strong correlation with this regional availability should appear. The action of mowing exposes a sudden flush of invertebrates and enables ground feeders to exploit fields they had previously ignored because of dense vegetation. The switch to silage that has taken place over the last 20 years (p. 158) has brought cutting forward into early to mid-May; a second cut is then possible in late June or early July. Many birds exploit this sudden rich source of food but this spacing and timing now fit quite neatly with the spacing and timing of Mistle Thrush broods in particular. Timing of breeding is apparently quite critical for Mistle Thrushes, for variation in their breeding success in a variety of habitats is correlated with habitat-specific variation in time of breeding (O'Connor, 1986).

Other ground feeders showed weaker relationships of this type (Table 3.4): the increase in rearing success in Blackbird was about one-third more in the south-west than in eastern England. Starling egg success improved

Fig. 3.3. Regional variation in (top) clutch size and (bottom) fledging success (percentage of eggs yielding fledglings) of Mistle Thrushes in relation to the percentage of the region's land area mown.

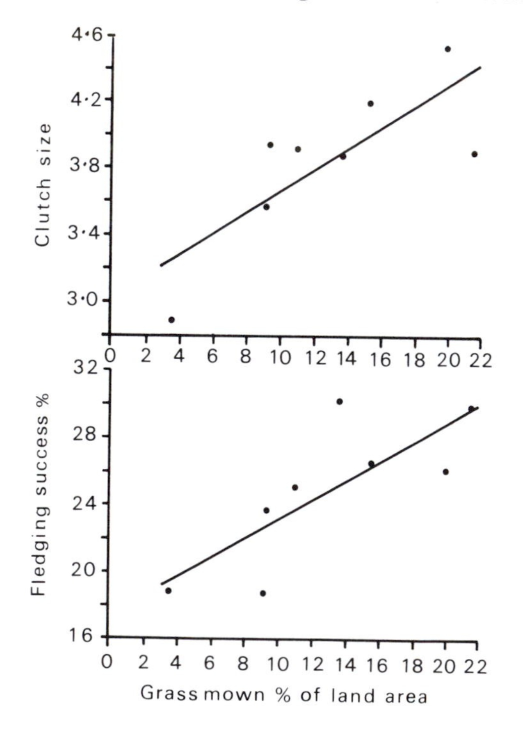

across the same gradient, and Chaffinch clutch size increased from an average of 3.8 to 4.4 eggs. However, in each of these species the other breeding parameters altered only erratically or not at all, so that none of them approached the Mistle Thrush in strength of response. No relationship between nest failure and mowing was apparent in the data for Whitethroat, although Mason (1976) found mowing to account for as much as 8 per cent of its nest losses.

Two other species – Lapwing and Yellow Wagtail – perhaps deserve mention in this context. Lapwing breeding success varied hardly at all with mowing, a point lending support to Murton's (1971) argument that its decline has been caused by changes in the carrying capacity of modern agricultural land: unless the land is good enough to achieve a rather fixed level of breeding performance, the birds do not settle there. One may then see population declines but not a decline in breeding success (p. 162). Our data for Yellow Wagtail were rather variable in quality but an apparent decline in clutch size was offset to some extent by improved rearing success in areas where much grass is mown. So, although about 15 per cent of nest failures in this species are due to agricultural operations such as mowing (Mason & Lyczynski, 1980), pairs that escape such losses otherwise do well when rearing young.

Breeding in grassland and in arable

Fig 3.4 compares laying periods in grassland and in arable habitats for Lapwings and for Skylarks. In grassland Lapwings show a sharply defined laying peak from late March to early April, and laying declines steadily thereafter. In arable land the main laying period extends over a much longer period and then stops rather abruptly. These differences seem to be caused by farm operations. In the grassland habitats used, management is relatively constant from year to year and Lapwings can start laying every year at about the same time. In contrast, in arable habitats the timing of farm operations varies from year to year and can influence laying: for example, territorial Lapwings may be pushed out of one field before they have begun to lay there, so that the nest cycle must start afresh in the new field. Farmwork also destroys nests, necessitating repeat laying, though nowadays such destruction is unlikely after mid-May. What is especially surprising, however, is that it is in grassland that the largest numbers of small (and possibly repeat) clutches of two or three eggs are found (Table 3.5) and Table 3.6 shows that egg survival is also considerably lower in grassland than in arable. Taken together, these factors argue a higher rate of predation in grassland, which we show elsewhere is not generally the preferred nesting habitat: tilled land is preferred, although intensive farming has pushed the majority of pairs into grassland.

The same analysis produced a similarly varied pattern in Skylarks (Fig. 3.4). For this species laying in grassland starts later than in arable, reaches a very sharp peak in early May and then declines very rapidly. This pattern

Fig. 3.4. Distribution of egg laying dates of (*a*) Lapwing and (*b*) Skylark nesting in (top) grassland and (bottom) arable. Data are from BTO nest record cards.

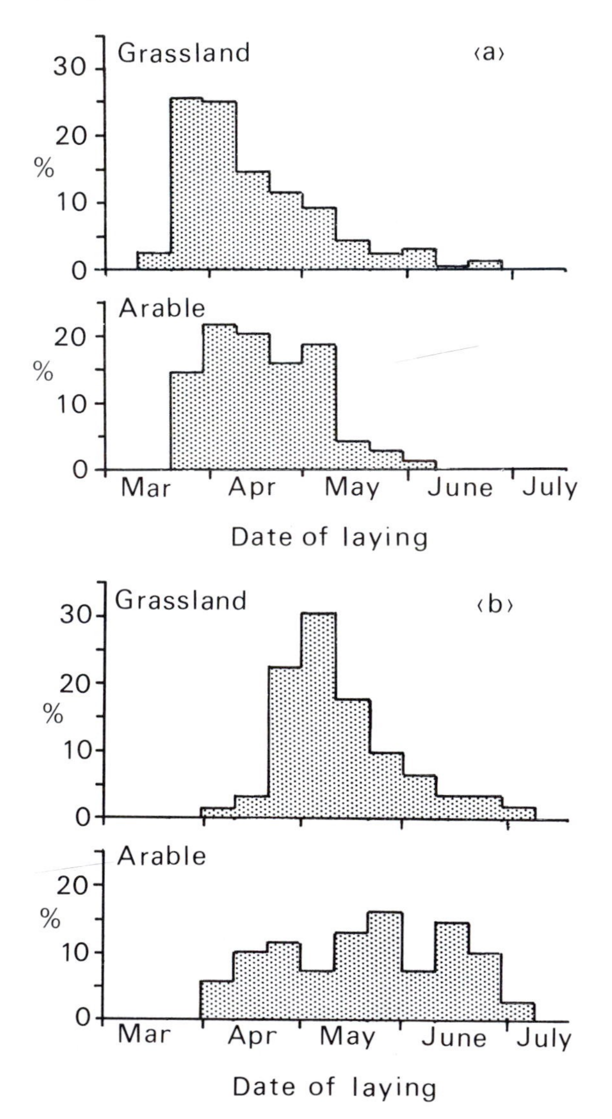

allows the birds to exploit the flush of invertebrate food now provided by the first cut of silage in mid-May. They probably then to a great extent leave the habitat since no peak corresponding to second clutches is apparent in the figure. In arable land laying starts earlier and the multiple peaks indicate multiple broods; the largest peak of laying coincides with second broads, so some pairs from grassland may move to arable to rear a second brood.

Examination of Tables 3.5 and 3.6 provides further information on these patterns for Skylark. In grassland, clutch size and egg and chick mortalities differ very little from the values observed in arable land but the nest record cards show that about 20 per cent more nests in grassland survive to produce some young. The chick mortality recorded in grassland thus derives more from partial losses from each brood whilst chick mortality in arable land occurs more as total brood losses. This means that the birds are exploiting the present pattern of grassland management to improve their chances of a successful nest at the first attempt. When examined over time, egg success in arable land has improved over the 20-year period studied. In Denmark Laursen (1981) found that Skylarks nesting around

Table 3.5. *Clutch sizes of various species in nests in grassland and in arable land in Britain, 1962–80*[a]

	Grassland		Arable land	
	Mean	s.d.	Mean	s.d.
Lapwing	3.62	0.69	3.77	0.54
Skylark	3.36	0.66	3.22	0.69
Dunnock	3.92	0.83	3.84	0.88
Blackbird	3.69	0.86	3.63	0.85
Song Thrush	4.12	0.67	4.10	0.80
Mistle Thrush	4.08	0.74	3.60	0.91
Whitethroat	4.79	0.70	4.52	0.70
Magpie	5.51	1.41	5.89	1.37
Jackdaw	4.13	1.07	4.38	1.12
Carrion Crow	3.75	1.14	4.00	1.41
Starling	4.69	0.93	4.93	0.88
Chaffinch	4.24	0.89	4.06	0.89
Goldfinch	5.09	0.61	4.92	0.76
Linnet	4.89	0.71	4.78	0.57
Yellowhammer	3.47	0.71	3.53	0.63
Reed Bunting	4.28	0.81	4.33	0.80

[a] Data are from BTO nest record cards.

areas of spring barley most often actually placed their nests in road verges. He suggested that the fields had too little cover and too much disturbance from farm operations for nesting at the start of the laying period, forcing them to use verges. It seems likely that this behaviour encourages predation, since it limits the area predators must search. Quite possibly, therefore, the change to winter rather than spring cereals has encouraged more nesting in fields and so has reduced predation.

A number of other species show differences in breeding biology between nests on grassland and nests on arable land. The Mistle Thrush produced slightly more eggs on grassland than on arable but clutch sizes were very similar in both habitats (Table 3.5). Breeding success was markedly more variable between habitats, particularly in respect of egg mortality (Table 3.6): Mistle Thrush, Whitethroat, Dunnock, Carrion Crow and Reed Bunting were each more successful with eggs in grassland areas. The first two were less successful there when rearing young, despite the advantage that mowing confers on Mistle Thrushes (see Fig 3.3); this is because only 21 per cent of grassland is mown for silage. In contrast, the Song Thrush

Table 3.6. *Egg and nestling mortality (per cent lost per day) for various species, in nests in grassland and in arable land in Britain, 1962–80*[a]

	Egg mortality				Nestling mortality			
	Grassland		Arable land		Grassland		Arable land	
	Mean	s.d.	Mean	s.d.	Mean	s.d.	Mean	s.d.
Lapwing	3.2	0.09	2.4	0.12	6.4	0.43	7.5	0.84
Skylark	6.5	0.35	6.3	0.39	8.3	0.50	8.6	0.59
Dunnock	4.2	0.30	6.1	0.43	4.8	0.41	5.3	0.60
Blackbird	4.1	0.32	3.8	0.41	5.3	0.44	5.9	0.60
Song Thrush	4.9	0.36	4.0	0.36	6.7	0.59	5.1	0.45
Mistle Thrush	4.0	0.32	5.3	0.74	5.6	0.43	4.3	0.80
Whitethroat	2.3	0.31	3.4	0.25	7.1	0.59	5.8	0.33
Magpie	4.0	0.19	3.5	0.26	2.4	0.20	2.5	0.33
Jackdaw	3.2	0.25	2.4	0.39	2.9	0.30	3.0	0.51
Carrion Crow	2.8	0.27	3.4	0.35	2.4	0.27	1.9	0.31
Starling	2.9	0.36	1.8	0.40	2.7	0.32	3.5	0.39
Chaffinch	4.5	0.30	4.6	0.34	6.4	0.47	6.4	0.53
Goldfinch	4.6	0.45	4.5	0.54	5.1	0.34	5.5	0.42
Linnet	3.4	0.25	3.4	0.28	5.1	0.34	5.5	0.42
Yellowhammer	4.9	0.34	5.0	0.29	2.5	0.42	4.4	0.65
Reed Bunting	3.5	0.30	4.7	0.54	6.8	0.53	6.2	0.42

[a] Data are from BTO nest record cards.

experienced greater losses of both eggs and chicks when on grassland. Starlings likewise lost more eggs there than on arable but then benefited from good nestling survival; this last was true also of Goldfinches. Whilst the details of these various differences are likely to be functions of individual species' ecology, it does seem that the main differences between the two habitats relate to predation (affecting egg mortality) and to finding enough food when rearing young (cf. the discussion of Mistle Thrush and Skylark above).

The influence of climate and altitude

Regional variation in farming practice, such as the incidence of grass mowing, is superimposed on patterns of climate and in some cases climate may be the factor determining breeding effort and success. Because spring is milder and earlier in the south and west of the country, one might expect various species of birds, if limited by spring temperatures, to start breeding earlier in such regions. For example, Chaffinches breeding in the south-west of the country lay a few days earlier than those nesting in the south-east and north of England and some two weeks ahead of birds breeding in Scotland (Newton, 1964a). Pied Wagtails and Yellow Wagtails show a similar pattern: 29 per cent of Pied Wagtail clutches were started in April in south-east England but only 24 and 20 per cent were started in northern England and in Scotland respectively: for Yellow Wagtail clutches, 19 per cent have started before 10 May in the south-east, against 9 per cent in the north (Mason & Lyczynski, 1980). The fact that Rooks have long been seen as a greater agricultural problem in Scotland than in England is closely related to these regional differences in breeding season. In Scotland the later sowing coincided with the birds' breeding season, when energy demands are large, thus promoting the consumption of seeds. In southern England sowing is more advanced than is the Rooks' season and the grain has grown away before the Rooks need it (Feare *et al.*, 1974).

Regional variations in relation to temperature may be more pronounced in resident species than in migrant species in which migratory concerns may synchronise breeding over the country as a whole. Thus Fuller & Glue (1977) found a two-week difference in the onset of breeding by resident Stonechats in the Channel Islands and in Scotland but found marked synchrony over the same latitudes by the migratory Whinchat. Newton (1964a) concluded that the annual variations in breeding season of the Chaffinch (variations extending over about a fortnight) are correlated with the prevailing temperatures, and the number of clutches begun in April and early May are correlated with the temperatures over the preceding five-day periods. When spring temperatures rise sharply, breeding activity also increases sharply and egg laying occurs more or less synchronously in the population as a whole. But when spring temperatures rise slowly, there is greater individual variation in the start of laying, and the egg-laying curve is broad and low.

Fig 3.5 shows how annual variation in mean first egg date of Pied Wagtails has varied in relation to mean April temperatures for England over the period 1955–76. On average a 1 °C rise in mean April temperature brought forward the mean date of first egg by approximately two days and corresponds to a 7 per cent increase in the number of clutches started in April. Since springs have tended to be colder in recent years such a relationship could generate secular effects. In fact, a smaller proportion of clutches have started each April in recent (1966–76) cooler years as compared to an earlier warmer period (1955–65) (Mason & Lyczynski, 1980). Lapwings, too, have been affected by the recent changes in climate (Jackson & Jackson, 1980). Since about 1974 the average rainfall between April and June in the Beaulieu area of Hampshire has fallen to about 42 per cent below the 1941–70 average, resulting in the heath drying out and providing conditions less suitable for newly hatched Lapwing chicks. Average hatching success decreased from 78 per cent during 1971–74 to 54 per cent 1976–78 and average chick mortality rose from 77 per cent to 94 per cent over the same period. Thus weather conditions are important in maintaining the right conditions for the chicks to feed adequately (though the Lapwings' problems on this particular site were aggravated

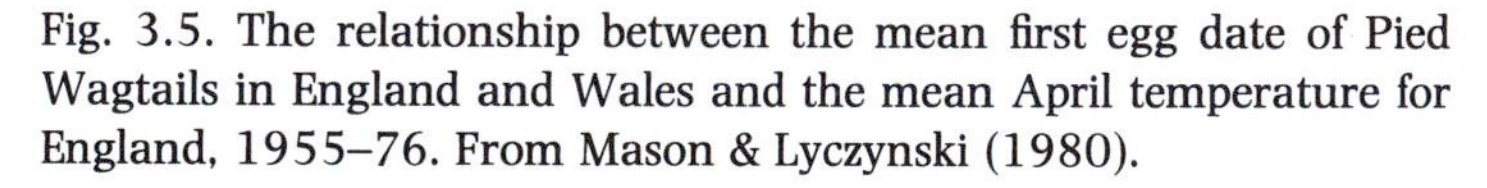

Fig. 3.5. The relationship between the mean first egg date of Pied Wagtails in England and Wales and the mean April temperature for England, 1955–76. From Mason & Lyczynski (1980).

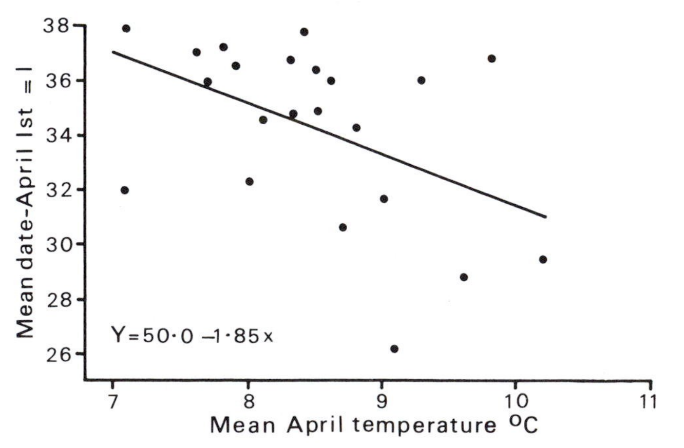

by the fine weather increasing the amount of human disturbance on the site).

The availability of food for laying females may be the ultimate cause of these correlations of breeding with temperature. In warmer weather a female spends less of her food intake on her own maintenance and so has more to divert to egg formation (Perrins, 1970). The results of 11 field experiments testing this theory have been summarised by Davies & Lundberg (1985). In all species but one (and including such common farmland birds as Dunnock, Magpie, Carrion Crow and Kestrel) the experimental provision of extra food resulted in earlier breeding. In the case of the Robin, breeding was not brought forward but the clutch laid was larger. Larger clutches were also laid by Magpies and Kestrels fed food supplements.

The breeding biology of birds resident on upland farms may be modified

Fig. 3.6. Distribution of laying dates for Curlew: (top) nesting in the lowlands below 152 m; and (bottom) nesting in the uplands above 152 m. From Morgan (1982).

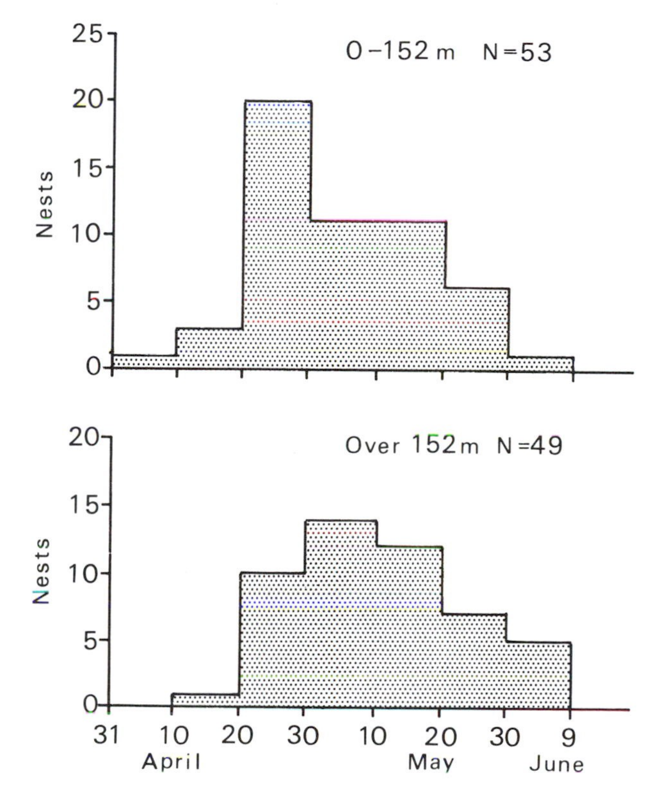

with altitude, as shown for Curlew in Fig. 3.6. Similarly, some 30 per cent of Pied Wagtails in lowland (below 30 m) in southern England have started laying by the end of April, against only 22 per cent of birds above 150 m. For Yellow Wagtails equivalent figures are 36 and 17 per cent started by 10 May (Mason & Lyczynski, 1980). A general feature of passerines nesting at higher altitudes is that the breeding season is more concentrated than in the lowlands. Meadow Pipits, for example, have laying lasting up to 130 days in different parts of lowland Britain but only 89 days at 200–400 m altitude and just 75 days above 400 m (Rose, 1982). For species such as the Meadow Pipit that use grassland habitat of very simple structure and that take predominantly surface arthropods in their diet, these altitude effects (and the similar pattern with respect to latitude that they also display) are possibly related to the effects of climate on the availability of their prey. Clutch size certainly decreases with altitude in the Meadow Pipit, as expected from this explanation (Rose, 1982). In contrast, among breeding waders probing species such as Curlew and Dunlin are more successful above 1000 feet (304 m) than they are in the lowlands below 500 feet (152 m); surface-feeding species such as Lapwing resemble the Meadow Pipit in their poorer success at the higher altitudes (Morgan, 1982).

For other species altitude may figure as one of the factors allowing coexistence between otherwise competing species. Whinchat and Stonechat segregate in this way, with 32 per cent of Whinchat nests located in agricultural habitat, primarily in the uplands, against only 2.3 per cent of Stonechat nests in such areas (Fuller & Glue, 1977). Two-thirds (67 per cent) of the Whinchat nests in agricultural situations were in grassland, and in Scotland were frequently concentrated where agriculture and moorland meet (Phillips, 1970). Only some 6 per cent of the nests were located in arable areas.

Summary

In summary, our data show that nesting patterns on farmland are by no means as clear cut as indicated by previous reviews. For most common hedgerow species predation rates are intermediate between those experienced in woodland and in suburbia, but nestling starvation is commoner than is generally acknowledged and varies between habitats in a species-specific manner. Breeding success varies regionally and for several species is closely linked with agricultural practice in space and time. Intraspecific variation in egg laying is associated with variation in average temperatures or in altitude, probably largely due to temperature-induced variation in food abundance or availability.

4: Food and feeding behaviour

Farms provide three major sources of food for birds. These are: (1) the invertebrates of the soil, crops and naturally occurring plants; (2) the green vegetation of the growing plants, mostly crops and grass; and (3) grain, seeds and hedgerow fruits. Four main factors influence the availability of these food supplies. First, relative abundance of foods may vary seasonally; in summer soil invertebrates are less readily found but seeds and grain peak at that time. Second, food availability varies with farm work, for instance when cultivations or mowing produce a temporary flush of easily found food items. Third, regional differences in farming practice may affect the distribution of food, notably of seeds of arable weeds. Finally, management may alter the availability of food, e.g. by varying the timing of cultivations, mowing or harvesting, or by changing the way such work is done. The distinction between availability and abundance is evident throughout this chapter. In farmland the first is often more immediately important to birds. For example, direct drilling techniques, which involve minimal cultivations, are known to increase the mass of soil invertebrates present (Edwards, 1984) but the lack of ploughing and tilling reduces the number of feeding opportunities available to birds.

The soil invertebrates most important as food for birds are the various species of earthworms, tipulid larvae (leatherjackets), beetles and molluscs (Murton, 1971). Earthworms are most abundant in April and May and from September to December. During the summer the dry conditions force many species to aestivate in deep burrows where they are inaccessible to birds, and in the late winter low soil temperatures also keep them well below ground. Leatherjackets similarly have a seasonal pattern, overwintering on grassland from August or September as small larvae, then growing rapidly from spring to July when they pupate, and emerging as adults in late summer. The combination of numbers and size results in peak biomass in May and June, a timing which in turn determines the breeding season of the Starling (Dunnet, 1955). Leatherjackets are most abundant in damp areas but pasture can hold considerable densities.

Important groups of surface-dwelling invertebrates taken by birds are the carabid, staphylinid and scarabeid beetles, some weevils, and the araneid spiders. Their biomass declines far less in summer than does that of soil dwellers such as earthworms and leatherjackets, and in some habitats increases substantially. Birds such as the Lapwing can therefore use such foods throughout the summer and immigrant Lapwings can come into Britain as early as July and find food in the field fauna, whereas specialists in soil invertebrates would starve (Murton 1971). Birds such as the Rook and Starling in fact have to switch diet in late summer, from probing for soil animals to feeding on surface-dwelling fauna (e.g. Macdonald, 1983). Immigrant populations of these species therefore await the autumn renewal of the soil fauna in October, remaining meanwhile near their breeding grounds where local knowledge is presumably valuable in locating some food. Against this late summer advantage enjoyed by surface-feeding birds is the greater vulnerability of the surface fauna to winter frosts, so that in some years surface-feeding birds may be more badly affected than probing species.

Although the commoner farmland species divide almost equally between species with largely animal diet and species with largely vegetable (or mixed) diets, many of the latter need some animal foods for the growth of their young and may be constrained by the availability of adequate invertebrate food for their nestlings. Subject to this constraint, seed-eaters are able to breed over a longer season than invertebrate feeders, but experience their worst conditions in late winter and early spring. Seeds and fruits develop largely during the late spring and summer and peak in summer and autumn, and the quantities available decrease steadily throughout the winter as the seeds are consumed, are broken down by weather, or germinate. Some farmland seed-eaters e.g. Linnets, therefore move out of Britain in the autumn if the seed supply is particularly low (Murton, 1971).

The winter community can make greater use of vegetable foods such as seeds and hedgerow fruits than can the breeding community, since the nutritional needs of growing young are no longer crucial. However, seeds are available only to those species able to search for them on the ground (Murton, 1971). There is little scope for a late summer influx of seed-eaters since the resident populations, already on site, generally breed at levels appropriate to the seed abundances present. Where immigrant seed-eaters do arrive, therefore, they often segregate ecologically from the residents. For example, Chaffinches resident in Britain are largely woodland birds taking a wide range of cereals and seeds, heavily supplemented by invertebrate foods. Immigrant birds are segregated from them by using open fields and by resorting to a restricted diet of cereals and their weeds, especially charlock *Sinapsis arvensis*, fat hen *Chenopodium album* and persicaria *Polygonum persicaria* (Newton, 1967). Agricultural practices in Britain may therefore have favoured these visitors until recently, when such weeds became the target of specific herbicides (p. 175).

One of the principal effects of modern rotations has been the concentration of cultivations in the autumn programme. Combine harvesting of cereals and the immediate cultivation or ploughing of stubbles, today increasingly followed by early planting, affects birds' food in two ways: (1) autumn food in stubbles is available for a shorter period; and (2) corn ricks and chaff heaps – once threshed through the winter – are no longer available as a food source. Autumn-sown crops also support different levels of surface and soil invertebrates than do spring-sown crops. Spring ploughing reduces the abundance of, for example, those carabid beetles that overwinter as adults, although species wintering as larvae deeper in the soil are less affected. Autumn-sown crops may therefore hold a richer invertebrate fauna in late summer and be more attractive as feeding grounds at that time than are fields of spring crops. But spring cultivations which bring fresh seed and invertebrates to the surface are a valuable food source for some birds at a time when other feeding opportunities may be scarce. Birds that benefit especially are seed-eaters and species taking soil invertebrates, such as thrushes, and these have declined since the recent shift from spring-sown cereals (Chapter 8).

Since the insect supply in any area may be closely linked to the diversity of the flora, the use of sprays almost routinely in arable farming can reduce the diversity of vegetation, not only in the cereal or root/vegetable fields but also, through spray drift, in adjacent hedgerows. Such effects have probably narrowed the range of invertebrates available to nesting birds for the feeding for their young, thus making them more vulnerable to natural hazards (Potts 1980).

Where bird populations are limited by food, closely related species tend to diverge in feeding habits (Lack, 1971). Hence a diverse community of species limited by food can exist only in a habitat which can offer a diversity of foods. Current agricultural practice in cereal farming and chemical use, however, tends to produce strong uniformity in vegetation, by techniques which are aimed at reducing competition from weeds and insect pests and at controlling disease. This is particularly the case in cereals which in 1970 accounted for about 50 per cent of the arable acreage, increasing to 61 per cent in 1983. Such crops can be used by birds in two ways, either as

Table 4.1. *Feeding positions of bird species (excluding gamebirds) observed in cereal fields on a Hampshire farm in 1984*

	Rotovated strip	Headland 0–6 m	Field crop 6–100 m	Field centre
Great Tit	(*)			
Wren	(*)			
Willow Warbler	(*)			
Jackdaw	(*)			
Robin	*	*		
Magpie	(*)	(*)		
Woodpigeon	*	*	(*)	
Song Thrush	*	*	(*)	
Blackbird	*	*	*	
Dunnock	*	*	*	
Chaffinch	*	*	*	
Linnet	(*)	(*)	(*)	
Mistle Thrush	*	*	*	
House Sparrow	*	*	*	(*)
Yellowhammer	*	*	*	(*)
Rook	(*)	(*)	*	*
Stock Dove				(*)
Skylark		(*)	*	*
Lapwing			(*)	*

A rotovated strip separated crops from field boundaries. Asterisks indicate the species was seen at that position. Parentheses indicate such sightings were rare. The three groups are of edge species, of species using the outer 100 m (but mostly 50 m) of the fields, and of species avoiding field edges.
From Fuller (1984).

hunting ground for invertebrate prey or as a source of grazing or seed. Table 4.1 summarises the results of a summer survey of crop use by birds on a Hampshire cereal farm and shows that few birds made much use of the fields. Such use as was made was concentrated in a rotovated strip bordering the fields and very few birds fed out in the centre of the fields. Most of the ground-feeding passerines present used the crops only near their edges, possibly because of the risk of predation, a pattern which Davis (1967) also found in his survey.

Such behaviour seems to be very variable and may reflect the range of species present and the food resources available. Fuller's study farm was rather poor in bird species; elsewhere Whitethroats, Reed Buntings, Yellowhammers, Sedge and Reed Warblers all forage regularly into standing corn in or around their territories and often into the middle of fields.

Crops as food

Although nearly every species inhabiting farmland takes crop plants or seeds occasionally, comparatively few birds feed to any great extent directly on the crops. Species for which crops form a large percentage of their diet include geese, swans sometimes, some ducks, gamebirds and rails, pigeons, Rooks and Jackdaws, and a few small passerines, mainly Skylarks, Greenfinches and sparrows. An important limitation for specialised grazers is the enormous quantity of green vegetation that needs to be consumed to provide sufficient nourishment. Species that feed on animal foods in summer and graze in winter have an additional problem, for the gut structure that is optimally adapted to process high energy invertebrate foods during the summer is less well adapted to process low energy vegetation during the winter: the gut must therefore alter in form each autumn and spring if the bird is to exploit vegetation in winter. Prŷs-Jones (1977) has shown for buntings that the intestine lengthens each autumn in parallel with the seasonal shift from invertebrate foods. Specialised grazers have a well-developed crop, a diverticulum from the oesophagus in which to hold ingested food until it can be digested, thus allowing feeding to proceed without constraint from the digestive processes.

The only species which relies extensively enough on crops to be a pest nationally is the Woodpigeon. Weed seeds are the principal food of Woodpigeons in early summer, followed by cereal grains in late summer and autumn, at first from standing corn but later from stubble fields. As the spilled grain is eaten or ploughed under, the birds rely increasingly on clover leaves and, to a lesser extent, on weed seeds and the leaves of winter corn and brassicas. Because clover does not grow during the winter, the birds steadily deplete the stocks and may starve (Murton, Isaacson & Westwood, 1966). Newton (1970) notes that the Cambridge–Suffolk population studied by Murton and his colleagues was therefore limited by the food supply, i.e. the winter stocks of clover.

Crop levels have also limited Woodpigeons in the Chichester area. Here birds feed in winter on brassicas (turnips, swedes and cabbages until the early 1970s, now oilseed rape), switching in early spring to clovers (both red clover crops and clovers established in pasture), and turning in summer and autumn to grain (including peas, field beans, standing cereal crops if these have lodged, and stubbles). Numbers present have fluctuated in line with cropping changes. Thus the size and number of winter flocks declined steadily with a decline in the area of turnips grown from about 1963–1975, and then showed a rapid increase with the spread of oilseed rape. Similarly, when grain maize was extensively tried in the area in the 1970s, a rapid but temporary increase in autumn numbers resulted, although numbers have otherwise usually been lowest in the autumn.

Geese also take significant quantities of crops and all species tend to be regarded, sometimes justifiably, as pests by farmers. Canada Geese have become a pest in the Midlands, where flocks of 400–600 can be found on leys, young corn, and potatoes ready for harvest. This species is a year-round resident but goose problems are also due to winter visitors. In the south and east of England Brent Geese have become something of a problem locally, grazing young corn, ley grass and pasture, a recent phenomenon associated with a rapid rise in total population. In the south the numbers feeding on crops or on grass tend to vary in line with the total present but it is by no means an exact correlation and there is an underlying trend for the habit to increase, once started, faster than the total wintering population. The birds are very fixed in habits, with strong preferences for certain fields, and they travel far from the estuary only in hard weather or when numbers are very high, or if they are harried. Of 20 counts made in the Chichester area over five winters, 75 per cent of birds feeding on fields were on grass, mainly pasture, and 25 per cent were on cereals, of which winter barley was possibly particularly favoured. Only well-grazed pasture is attractive to them. In Essex the most serious problems arise in late winter and spring, particularly during periods of wet weather (Deans, 1979). The combination of grazing and puddling covers the growing shoots with soil which then hardens and inhibits plant growth. Brent feed in particularly dense packs and are very persistent once a pattern of feeding is established. This may take as little as three days.

In contrast, grey geese take nearly all their food from farmland sites and range much more widely from roosting areas (Owen, 1980). White-fronted Geese probably take crops least, apparently preferring wet grassland. Greylag and Pink-footed Geese regularly graze cereals, scavenge potatoes after the harvest, and sometimes take other crops such as carrots, turnip leaves and other brassicas. As with Brent Geese, numbers have increased sharply in recent years.

The Stock Dove spread throughout Britain as cereal farming increased (p. 4), though it probably evolved in a parkland ecotone. It specialises on the weed and grass seeds associated with arable farming, particularly *Polygonum*, and, until the advent of organochlorine seed-dressings in the early 1950s, was a significant member of the bird community of such farms. Its numbers declined sharply during the organochlorine era, when the population contracted into coastal and less intensively arable areas (O'Connor & Mead, 1984). It subsequently recovered when the chemicals concerned were withdrawn from use as seed dressings.

Collared Doves apparently feed almost entirely on grain or grain products (e.g. bread) and seem to be totally dependent on human activities. Murton (1971) has in fact suggested that the reason for the success of the Collared Dove in Britain and Europe has been the existence of a niche for a small dove commensal with man left vacant by the decline in popularity of the dovecote pigeon. Collared Doves may prefer buildings and the like because they have a tradition of being semi-domesticated in the ancestral Near East, though what triggered their explosive spread into Europe is unknown (Hudson, 1965). Maize fields may at times attract sizable flocks but the principal habitats in use are around grain and seed stores and mills, and around intensively kept stock fed on grain, whole maize silage, so on, though in towns food provided on bird tables is frequently taken. An interesting habit, which shows how quickly they learn, is the way they work the country roads in the harvest, picking up the grain which trickles from the grain trailers as they go back and forth to the drier. Stubble feeding by this species is, however, rather uncommon.

All finches, sparrows and buntings take some grain but it seems an important food source only to House Sparrows and Greenfinches, which can exploit standing cereals as well as waste grain, and to Yellowhammers, particularly where farm stock are grain-fed (p. 67). The Skylark is the only other small passerine to take much grain. Few small passerines appear to take crop plants to any extent. Skylarks have frequently been reported to be grazers of young seedling plants, locally causing severe damage to such early season cash crops as lettuce and peas, and also causing damage to winter wheat (Murton, 1971; Green, 1978). They also graze brassicas in severe frost or snow. Skylarks are also the most important predators of sugarbeet seedlings (Green, 1980). Densities of Skylarks on sugarbeet seedlings are high in fields with abundant seedlings, irrespective of whether these are of weed plants (the preferred food) or of sugarbeet. But where the sugarbeet field is particularly weedy Skylarks do not graze the crop but concentrate on the weeds (or on their seeds and any invertebrates, especially beetles, present). Where weeds are scarce amongst the sugarbeet – as a result of herbicide applications, for example – the birds concentrate on the latter and may inflict heavy damage, particularly when the plant is initially sown at the 6-inch or 9-inch spacings desired (the normal practice today). Formerly the plants were sown thickly and thinned to these

Table 4.2. *Foods taken by farmland Corvidae*

	Carrion Crow	Rook	Jackdaw	Magpie
Potatoes	—	+	—	—
Root crops	—	+		
Peas and beans	—	+	+	—
Grain	×	×	×	×
Wild plant seeds	+	+	×	+
Top fruit	—	—	—	—
Wild fruits and soft fruits		—	—	—
Nuts	—	+	—	—
Sick sheep and lambs	+	—	—	—
Live mammals	×	—	—	×
Live birds	+	—	—	—
Carrion	×	+	—	+
Birds eggs	+	+	+	×
Nestlings	+	—	+	+
Other vertebrates	+	—	+	—
Earthworms	×	×	+	—
Terrestrial molluscs	+	+	+	+
Grassland insects	×	×	×	×
Woodland insects	—	+	+	+
Spiders, etc.	—	+	+	+
Ticks			+	+
Other invertebrates	+	+	—	—
Bread	+	+	+	+
Animal feed-stuffs	+	—	+	—

× an important food on farmland; + frequently recorded; — infrequently recorded.
Based on Holyoak (1968).

spacings by hand-hoeing the seedling rows. Modern methods of sugarbeet growing involve minimal hoeing and therefore rely heavily on herbicides, the use of which has increased approximately four-fold between 1969 and 1982. The emergence of the Skylark as a pest in sugarbeet can thus be regarded as a direct result of the greater efficiency of modern methods in eliminating weeds (Jepson & Green, 1983).

Holyoak (1968) has reviewed the foods of British corvids. Table 4.2 summarises the general patterns, particularly in respect of farmland foods. Grain is an important food of Carrion Crows, Rooks, Jackdaws and Magpies, being taken at all times of year but especially in autumn and winter – in the autumn usually from stubbles but in the later part of the winter and early spring mainly from sowings. Some grain is taken from standing crops in late summer and at all times of the year some may be taken from animal feeding troughs. Rooks and Jackdaws take some farm crops such as peas and beans in the autumn and winter, and during hard weather potatoes and root crops are taken by Rooks, Jackdaws and Crows. In spring and summer most of the crow species, but particularly Jackdaw and Magpie, take large numbers of grassland insects. Earthworms are an important food for Rooks throughout the year and from autumn through to spring are also commonly taken by crows. The other crows have beaks that are too short for them to reach many earthworms.

Other species take crops in small quantities; examples include swans, surface-feeding ducks, Coots, Moorhens, Pheasants and the two partridges, all of which graze cereals and some of which, e.g. swans, Mallard and Pheasant, also take other crops. Damage by Coots, Moorhens and Pheasants has been serious where these species are numerous but such behaviour is largely local opportunism. In summary, therefore, although many species take crops in small quantities only geese, Woodpigeons, Stock and Collared Doves, sparrows, Greenfinches, Rooks and Jackdaws use them as a major food source. This source may be important for only part of the year but can still help to maintain numbers, as discussed elsewhere for the Rook (p. 165).

The influence of farm animals

The presence of cattle and sheep on farms is beneficial to birds. Firstly, stock needs grassland, which much improves the feeding opportunities available to species that feed on soil invertebrates, notably Lapwing and Starling, corvids, and the thrushes. For these species grass provides a valuable feeding area when the growth of crops has made feeding on arable land difficult or impossible, and is also the preferred feeding habitat in winter.

Common farmland raptors, such as Kestrel and Barn Owl (and Short-eared Owl in winter), also rely extensively on grassland habitat for hunting (Shrubb, 1980; Pettifor, 1983) and various types of grassland provide the preferred habitat of certain specialised birds such as Snipe, or Wheatear (p. 160). Secondly, the presence of cattle and sheep ensures that water is always available, no matter how dry or how cold the weather may be. Water troughs are often difficult for small birds to drink from but cattle particularly are messy drinkers and leave puddles that birds can use. This may be important in severe winters, for freezing conditions deprive birds of water.

Thirdly, the places where stock are fed provide feeding areas for many species and may be particularly valuable in winter. Thus at Oakhurst in 1981–82 some 35 per cent of all small passerines noted on regular fixed walks were feeding around buildings or stock-yards in November, increasing to 66 per cent in a cold December, then falling in the post-Christmas period when there was a total thaw. In the blizzard of early January 1982 numbers around the buildings and stock-yards increased steeply to 96 per cent. Spencer (1982) has reported similar observations by BTO staff from farms in Hertfordshire during the snow of 1978–79. The species were Blackbirds, Chaffinches, Starlings, Dunnocks, Robins,

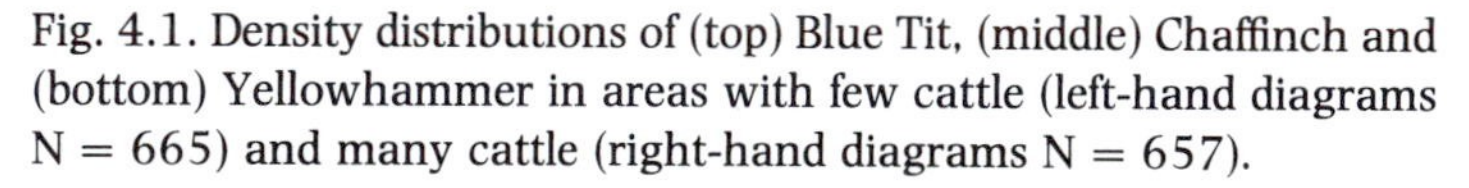

Fig. 4.1. Density distributions of (top) Blue Tit, (middle) Chaffinch and (bottom) Yellowhammer in areas with few cattle (left-hand diagrams N = 665) and many cattle (right-hand diagrams N = 657).

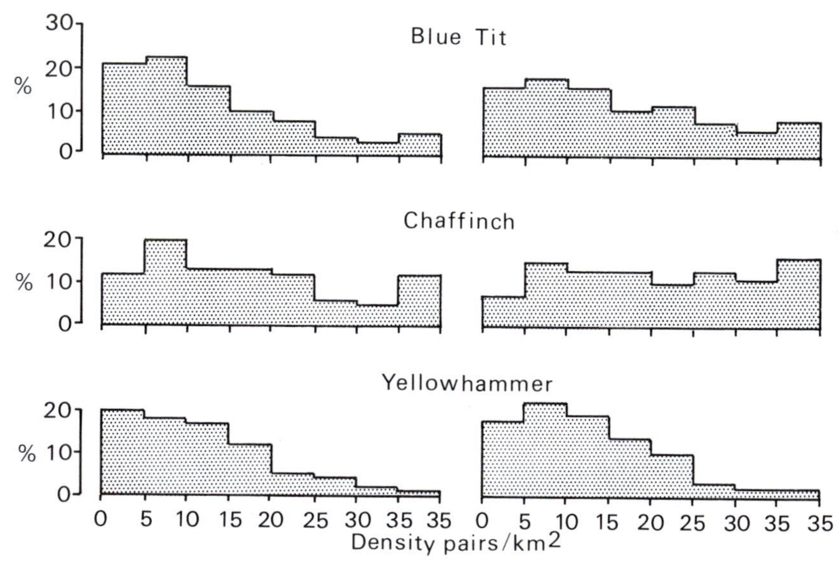

Song Thrush, Blue Tits (probably the entire breeding population of the farm), sparrows and Yellowhammers. Such birds exploit the feeding of grain to cattle, a practice which has in general increased with the decline in root crops. An interesting technical point is that cattle feed is now more often of whole-crushed grain rather than of grain ground into meal. The exploitation of this food source at Oakhurst, particularly by Yellowhammers, dates from this change.

Some species regularly winter around stock-yards, not leaving until the

Fig. 4.2. Blackbird densities in relation to cattle rearing in different regions of England. Within each region the upper diagram is for areas supporting few cattle, the lower is for areas with many cattle. E, Eastern England; SE, South-east England: EM, East Midlands; WM, West Midlands; SW, South-west England.

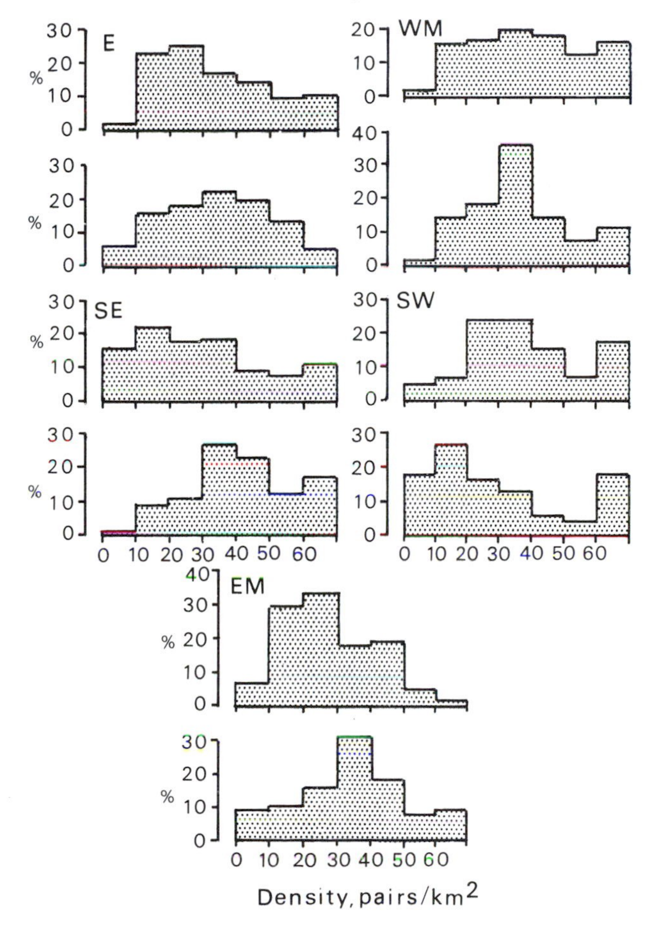

stock is finally turned out. Fig. 4.1 shows that over the period of the Common Birds Census densities of Blue Tits, Chaffinches and Yellowhammers (all of them species that regularly frequent cattle-yards in winter) have been systematically higher in cattle-rearing counties than in counties with low numbers of cattle, despite the many mild winters during this period. Several other species examined in this way showed such correlation with cattle numbers within some but not all MAFF regions considered. Because of other differences in farming practice the comparisons could be made only within five of the MAFF regions – South-east England, East England, East Midlands, West Midlands, and South-west England. The results for the Blackbird are presented in Fig. 4.2 and show a well-marked tendency for the association with cattle to promote Blackbird densities to a greater extent the colder the region is in winter: compare, for example, the South-east or East Midlands with the West Midlands or South-west. This was a general pattern, for when we did such analyses for 10 species (those mentioned already plus Skylark, Song Thrush, Dunnock, Starling, Goldfinch and Linnet), the association was most frequent in the South-east: only the Skylark (which nowhere showed any effect) and the Starling failed to show an association with cattle there.

Stock may also alter the habitat they graze in ways that change the prey spectrum available to birds, either by changing the animal community therein or by making the prey more vulnerable to bird predation. Table 4.3 shows how the diet of a Barn Owl in Surrey altered when an area changed

Table 4.3. *The effects of grazing on the prey taken by Barn Owl in a field in Surrey*

	Ungrazed		Grazed	
	Prey items taken	% of total weight	Prey items taken	% of total weight
Short-tailed vole	82	60	71	35
Wood mouse	24	18	49	24
House mouse	12	9	5	3
Bank vole	9	7	41	20
Common shrew	12	4	26	6
Pigmy shrew	9	1	16	2
Water shrew	2	1	1	1
Harvest mouse	1	1	11	2
Brown rat	0	0	5	7

From Glue (1967).

from a wet meadow overgrown with a thick mat of long grass in 1961 to a short grass pasture in 1964, following grazing by horses (Glue, 1967). After the disappearance of long grass the area became unsuitable for short-tailed voles but was used by brown rats, bank voles and wood mice from adjacent woodlands. Grazing stock are attractive to certain species which feed on the invertebrates disturbed by the stock. Hirundines take a lot of food around the feet and noses of grazing cattle and Yellow Wagtails regularly feed among them and sheep. Starling flocks also follow grazing cattle in winter. It is easy to underestimate the value of a food source like this. Some birds also feed on the invertebrates attracted to cowpats. Finally, during periods of severe weather stock often break through the snow cover by trampling in the course of their own feeding, thus allowing birds to gain access to food they could not otherwise use (Spencer, 1982).

The exploitation of farming operations

Almost all farmland birds make use of the feeding opportunities provided by farming operations. Beneficial operations take two basic forms – tilling the soil, which exposes soil invertebrates and brings seeds to the surface, and harvesting, which suddenly clears dense stands of herbage such as cereals and grass for hay or silage, and exposes invertebrates, small mammals, and seeds. The harvesting of potatoes and other root crops both clears the herbage and disturbs the soil.

There is a well-marked seasonal progression of farm operations through the sequence: spring cultivation – hay/silage harvest – cereal harvest – autumn clearing (cleaning) operations and root/potato harvest – autumn sowing – winter ploughing. Timing varies with soil and region, for the north is later than the south and spring cultivations on warm soils such as chalk are up to a month earlier than on heavier soil in the same area; the duration of the season also varies according to the type of soil. Mixed farms provide the best spread of such feeding opportunities and specialised cereal farms provide least, as the latter tend to provide little activity in the spring and early summer. But in any comparatively small area some work is likely to be going on which provides feeding chances for birds prepared to move about and many species appear to do so. This is especially true of the larger species; Kestrels and Barn Owls range over areas of up to 1000 acres (247 ha) and such species as Rook, Jackdaw and Stock Dove range over several different farms even during the breeding season. Black-headed Gulls in Wales will travel several miles from their hill nesting pools, and Mistle Thrushes will move up to three-quarters of a mile from nests, to feed behind the plough or on recently cut grass. Blackbirds and Song Thrushes will also range well outside their normal feeding territories to exploit freshly tilled fields; birds of several pairs will gather food in the same field where this cannot be defended as territory by any one pair. Such a wide-ranging search for food is not strictly confined to the exploitation of farm work. Twite, for example, will move up to 2 miles from moorland nests to gather seeds from pasture on the moorland fringe (F. Gribble, personal communication).

Because farm work is seasonal and well worth exploiting, many farmland birds have clearly defined seasonal feeding patterns. Birds have definitely learned to associate farm work with food, and species such as Kestrels, Rooks and gulls will investigate a tractor working in a field no matter what it is doing – spraying, for example. In this way an annual cycle of feeding behaviour is locked into the cycle of feeding opportunities generated by the farming year and will reflect any changes in that year brought about by new agricultural practices. The result of all farming systems is that potential food stocks for birds are at their lowest between March and May. All the stubbles are finally ploughed, all the hedges have been trimmed and the cuttings burned, and all the ricks and chaff heaps and so on have finally been disposed of. The food source provided by tillage bringing fresh seed (and invertebrates) to the surface may therefore be vital to potential breeding birds. But the proportion of spring tillage has declined drastically as the proportion of autumn cereals has increased and that of root crops and summer fallows declined (Chapter 1). The impact of this change is dicussed in more detailed in Chapter 8.

Other localised feeding

Glue (1967) studied the diet of Barn Owls in different farmland habitats. His results are summarised in Table 4.4, which shows how prey taken differed substantially between habitats. The differences shown between rough and wet pasture and parkland and grazed meadows also generalise the observations in Table 4.3. Short-tailed voles in grassland typically undergo population cycles over periods of 3–4 years and in the years of scarcity the owls changed hunting grounds to prey primarily on brown rats on mixed farmland. In areas of intensive cereal production practically all hunting is located along hedges and ditches, with as much as 25 km of such habitat needed for successful breeding. Kestrels show similarly localised hunting and are thus also susceptible to management changes involving hedge and ditch clearance (Shrubb, 1980). On this point we differ from Bunn, Warburton & Wilson (1983) who concluded that changes in agricultural practice were unlikely to affect Barn Owls.

Some species exist in farmland without greatly exploiting the opportunities presented by farming. Their presence is set by the structure of the farm rather than by its management. Thus, Wrens seem virtually confined to hedgerows, gardens and scrub. Great Tits, far more than Blue Tits, seem confined to trees and large shrubs and do not visit stock-yards in winter except in very severe conditions. Dunnocks feed far more around hedges and ditches than in fields, although they do move out into the latter. Tawny and Little Owls must have hunting perches and in farmland these are primarily provided by hedgerow trees and by buildings.

Ditches can be very valuable for feeding in winter. In farmland in Northern Ireland, Moles (1975) found that the relative use of wet ditches was nearly 50 per cent higher in winter than in summer, most markedly in the cases of Blackbird, Reed Bunting and Snipe. Robins overall made rather little use of wet ditches but, even so, this use doubled in winter. In contrast, Song Thrushes used ditches extensively in summer, perhaps because they provided good feeding areas in mid-summer, when the soil fauna has moved deep to avoid desiccation. In England Snipe feed in arable field ditches in hard frost but not otherwise. Thrushes, Robins and Dunnocks feed there in the same conditions but the thrushes tend to move out as soon as conditions ease. Robins are territorial in winter and so are rather restricted spatially but even so appear to confine themselves by preference to areas of trees and large shrubs at all times. Comparison

Table 4.4. *Prey taken by Barn Owls in different farmland habitats*

	Per cent of prey weights				
Prey species	Rough grass and scrub	Wet pasture	Mixed farmland	Parkland	Grazed meadows beside woods
Short-tailed vole	77	53	40	30	28
Rat	2	8	26	29	19
Shrews	6	8	12	11	24
Wood mouse	4	13	9	22	14
Bank vole	1	9	2	5	10
Others	10	9	11	3	5

Data from Glue (1967).

of the summer and winter habitat preferences of the Robin on Irish farmland shows only rather minor habitat shifts in winter, when tall scrub and pasture are visited but tree use declines (Moles, 1975). Spencer (1982) has drawn attention to the importance of molehills in providing freshly turned soil for this species. Like several of the others mentioned above (as well as other more specialised birds), the Robin seems to exist in farmland in spite of rather than by exploiting farming practices. This may be a feature of the relatively mild climate of Britain, for elsewhere many of these same species are heavily dependent on human activities. Thus Hilden & Koskimies (1969) concluded from their study of the effects in Finland of the very cold and snowy winter of 1965–66 that Great Tit, House Sparrow and several finches and corvids were entirely dependent on human habitations for their continuing survival as wintering birds.

Winter finch flocks seem particularly associated with four farmland habitats – chaff heaps and stack-yards, cereal stubbles, first winter leys where these are established under a cereal nurse, and unsprayed kale crops. The distribution of such feeding sites is very variable and depends, on any farm, on the rotation practised, on the form and standard of management, and probably on the distribution of food plants. The volume of food formerly available to seed-eaters could be very large. Seed production figures given by Salisbury (1961) indicate, for example, that an infestation of 10 plants/m^2 of fat hen, which is in no way improbable, may yield about 242 kg of seed per hectare of roots or spring cereals. This figure may be viewed against the estimates of annual biomass consumption by sparrows of between 2.87 and 4.00 kg dry weight per ha per year, of which up to 3.1 kg could be cereal grain (Wiens & Dyer, 1977).

The flocking habit of finches and buntings provides the most efficient way of finding and utilising the scattered concentrations of such food supplies. A large flock is more likely to locate the food source than is an individual searching solitarily and the food source once found can support many birds. In southern England the main constituents of feeding flocks are Chaffinches, Greenfinches, Linnets, sparrows, and Reed Buntings. Corn Buntings and Yellowhammers are nearly always likely to concentrate at the same food sources but tend to feed in more scattered groups and parties, and to be underrecorded. Such flocks may represent concentrations of both local residents and winter visitors.

During the past 25 years radical changes in farming systems and methods have extensively reduced the feeding sites for such flocks. The most significant and rapid changes took place between 1948 and 1959, with the disappearance of chaff heaps and corn ricks as the combine harvester replaced the self-binder and threshing drum (Table 5.1 below). The widespread introduction of herbicides to control weeds then reduced the abundance of seeds available to seed-eaters. The change in harvesting methods is of particular relevance to winter finch flocks. With the self-binder, everything was cut, stooked, carted and ricked, and then threshed. Both grain and weed seeds dried out in the stook and each handling caused some to be shed, so that handling spread seeds. When grain was threshed the chaff, light grains and weed seeds were blown into a separate heap, an operation which provided a continuous supply of fresh feeding sites as the threshing machine moved around the farm. This was usually a winter job and actually preserved a food supply for the seed-eaters. The combine harvester produces quite a different pattern. Everything is left until the grain is ripe, then the machine cuts the corn and separates the grain. Weed seeds, etc., are blown out of the back, where they fall under the straw. If this is baled, the seed remains available to birds, but if the straw is burned, much seed may be destroyed.

Finches are very mobile in winter and the decline in food sources can result in more extensive movements and greater concentration of flocks. The Winter Atlas shows the Midlands to be the region of greatest concentration of wintering finches and buntings, so the systematic records

Table 4.5. *Winter flock sizes for various species in the West Midlands*

Species	Typical range of flock sizes	Normal maxima	Status
House Sparrow	Up to 700	1200	
Tree Sparrow	50–150	1000–1500	Stable flock size but more flocks recorded
Chaffinch	100–200	500 (–1000)	Increased in size by one-third since 1973
Brambling	Less than 50	300 (–2000)	Fluctuates irregularly
Greenfinch	50–250	750	Stable flock size
Goldfinch	Up to 200	300–600	Maximum size increased from *c.* 100 in 1950s and early 1960s
Linnet	Mean of 200	500–1200	Increased in size and frequency to 1969–73; slight decline since.
Yellowhammer	Mean of 80	100–350	
Reed Bunting	Mean of 60	100–300	
Corn Bunting	Small	60–100	Increase in roost sizes

Data from Harrison *et al.* (1982).

of flock size kept by the West Midlands Bird Club are of particular interest (Table 4.5). They show that Chaffinch, Goldfinch, Corn Bunting, and Linnet flocks increased until the early 1970s. The increased size of Corn Bunting roosts is probably associated with its numerical increase in the area rather than with changes in conditions for winter feeding. This may also be so with the Goldfinch. But such changes have not been universal and in Sussex, for example, winter finch flocks have virtually disappeared since the mid-1970s. On the West Sussex Downs this has been attributed to the reduction in weed seeds in kale crops sprayed with herbicides but on the coastal plain of Sussex these flocks largely disappeared much earlier, having become rare by 1965.

In general, sparrows and Greenfinches have been the least affected species as they make greater use of cereal grains which are now more readily available than other foods. This probably explains the present dominance of Greenfinches in winter flocks. The Greenfinch has also switched to feeding at suburban bird feeders (Glue, 1982). Reed Buntings have shown a similar tendency recently and the increased extent to which Chaffinches and Yellowhammers are becoming birds of farmsteads in winter may be part of the same process.

Seasonal changes in feeding behaviour

Autumn and winter feeding habitats vary regionally. In coastal Sussex Lapwings feed primarily on arable fields. After harvest they are particularly fond of feeding over burnt stubbles. They follow the drill round the farm during autumn sowing, concentrating on the most recently tilled land, and then drift back over the young corn. In Buckinghamshire and in Bedfordshire, however, much winter feeding is on grassland (Fuller & Youngman, 1979; and personal observations). On both arable and grassland Lapwings show strong preferences for particular fields in any winter and on grassland may return annually to traditional fields. On arable, of course, cropping patterns may change from year to year, thus interrupting the possibility of such return. Winter wheat following a clover ley or other green crop, e.g. oilseed rape, seems to be a favoured arable crop in early and mid-winter but the birds may shift back to grass from mid-February on. Newly ploughed fields are always an attraction, however, and ploughed fields are also favoured for roosting. Most of the birds in Britain in autumn and winter are visitors, so it is difficult to know whether these seasonal and habitat shifts are by the same individuals or by different populations.

The regional differences in behaviour noted above may well relate to local conditions, especially climate. Although Lapwings use traditional feeding areas year after year, presumably because they afford reliable sources of invertebrate food, they are also rather opportunistic feeders, readily exploiting temporarily flooded grassland or pasture that has been spread with slurry or manure. All three events probably bring invertebrates nearer the surface, either to avoid drowning or to feed on the fresh organic material. Lapwing are also more flexible than, for example, Golden Plovers and will more quickly move away in response to frost and snow (Fuller & Youngman, 1979). However, there is a marked north–south variation in this behaviour, with proportionately more birds leaving Scotland and northern England than southern England in such conditions (S. R. Baillie, personal communication). During the hard weather of December 1981 and January 1982 some 79 per cent of all feeding records at Oakhurst were on arable but the birds switched to grass in frost and shifted straight back as soon as there was any thaw.

Gulls feed over farmland from August to April, the species varying regionally. They feed extensively on burnt stubbles after harvest, some of which may be favoured for two or three weeks, suggesting that this behaviour is more than a response to the immediate effects of burning. Spring and autumn cultivations again attract them but during the winter most are found on grassland (Vernon, 1970), although some Black-headed Gulls regularly feed over fields of young cereals, steadily quartering them in flight-hunting. The most numerous species are Common and Black-headed Gulls and Herring Gulls; Lesser Black-backed Gulls make rather little use of farmland except on passage, when many follow the plough. Common and Black-headed Gulls show marked altitudinal and habitat differences in distribution (Vernon, 1970). Some 65 per cent of Common Gull flocks found inland in winter were on grass, against only 37 per cent of the Black-headed Gull flocks. In addition, Black-headed Gulls were predominantly lowland species whilst 42 per cent of Common Gulls recorded were seen above 300 feet (91 m), largely because of a preference for drier and better drained grassland. Lowland records of Common Gull are primarily on well-drained and well-managed grazed pastures, such as the Romney Marshes of Kent, the Holderness area of East Riding, and the machair of the Outer Hebrides. They are largely absent from Midland counties of heavy clay and poor drainage, even though many are above 200 feet (61 m).

Blackbirds and Song Thrushes feed quite extensively over fields of young corn in winter. Thus in November and December 1981 and January 1982 36 per cent of feeding Blackbirds and 26 per cent of feeding Song Thrushes noted at Oakhurst were on arable land, largely young corn. The extent to which thrushes feed in cereal fields after the early spring seems to depend on the availability of alternative feeding habitat. Fuller (1984) found that Blackbirds nesting in the small woods at Manydown Farm in Hampshire rarely used the cereal fields but those nesting in the rather poor hedges on the farm made extensive use of the crops in summer. The woodland birds probably had sufficient food available to them without using the poor feeding conditions in the crops. The birds fed more on the initially bare and later weedy rotovated strips around the field boundaries than they did on the crops (Fig. 4.3). Fuller found little difference in the extent to which birds fed over spring barley, winter barley, and wheat. Grassland, particularly permanent pasture, is unlike cereals in being important to thrushes throughout the year. The bulk of breeding birds on farmland are around pasture (or gardens with lawns) and, in the winter months at Oakhurst, about 66 per cent of Song Thrushes, 36 per cent of Blackbirds and 82 per cent of Fieldfares and Redwings were noted feeding in pasture. In County Down in Northern Ireland pasture was the most frequented habitat used by Blackbird and Song Thrush but the immigrant thrushes preferred bushes and hedges (Moles, 1975). In England feeding by immigrants is seasonal, the birds moving onto pasture once woodland and hedgerow berries are exhausted. In summer many birds of the year (of Song Thrush and Blackbird) may disperse into crops such as clover after they

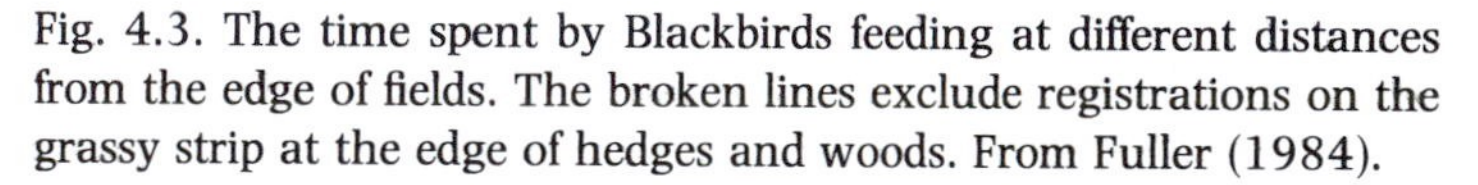

Fig. 4.3. The time spent by Blackbirds feeding at different distances from the edge of fields. The broken lines exclude registrations on the grassy strip at the edge of hedges and woods. From Fuller (1984).

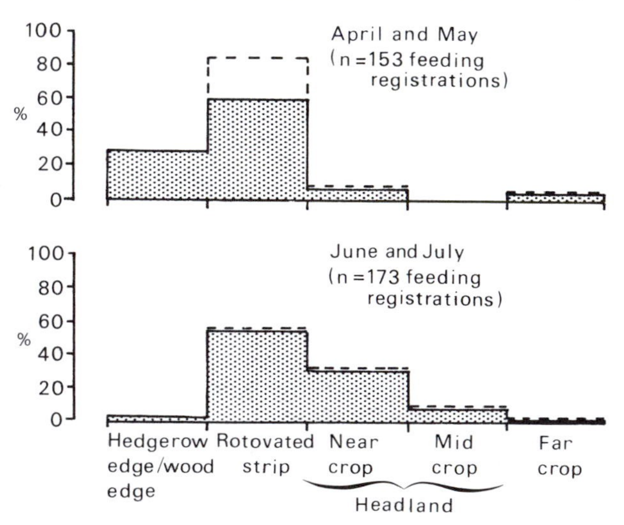

have fledged and may live there for an extensive period. They can be flushed all over the field but otherwise Song Thrushes and Blackbirds tend to work a band around the field not too far away (*c.* 50 m) from the hedge unless there are special circumstances, such as cultivations. The effect is clearly apparent for cereals in Fig. 4.3.

Although Mistle Thrushes work cultivations in spring (see above), they switch to grassland in early summer (p. 48). In autumn and winter they depend heavily on berries and will defend good bushes or trees against competitors, both conspecific and interspecific (Spencer, 1982; Snow & Snow, 1984). In contrast, the Winter Atlas results show that in winter Song Thrushes move to coastal areas over much of Britain; only the southern part of the country retains birds inland (Lack, 1986). Where they remain, Song Thrushes seem to change their habitat in winter. Fig. 4.4 shows this habitat shift on the County Down farmland studied by Moles (1975): habitat use in winter is less restricted than in summer. Song Thrushes at Oakhurst similarly move away from the vicinity of buildings and gardens in winter to a greater extent than Blackbirds: only 7 per cent of

Fig. 4.4. Spectrum of Song Thrush feeding habitats in County Down, Northern Ireland, in (top) summer and (bottom) winter. The index of preference is a measure of relative use of different habitats, corrected for availability. Data from Moles (1975).

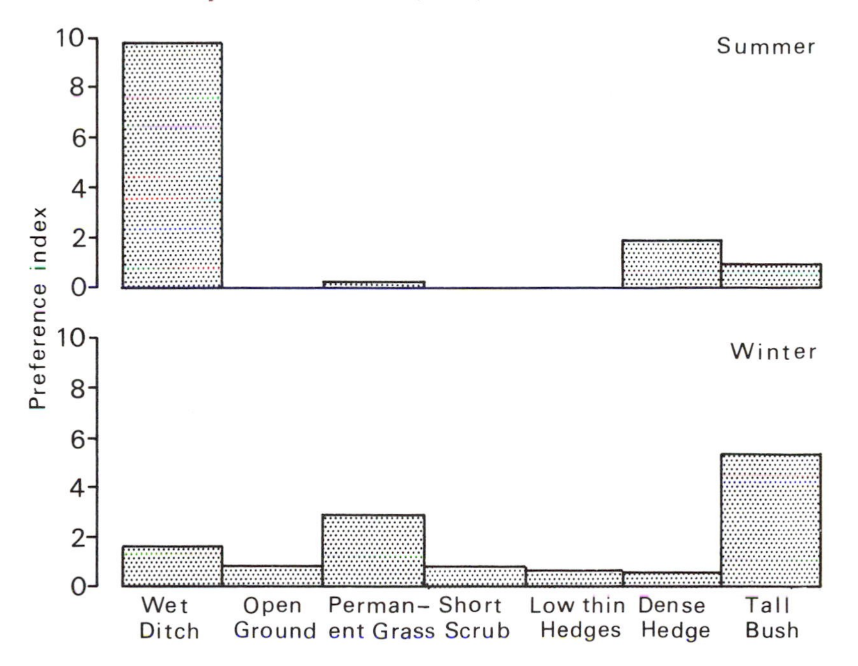

the winter Song Thrush records there were for buildings, stock-yards, and the like, compared to 30 per cent for Blackbirds. The latter are probably dominant to Song Thrushes in such situations. Ringing returns have shown a decrease since the turn of the century in the ratio of Song Thrushes to Blackbirds in Britain (Spencer, 1983).

Summary

Farmland provides three basic food resources for birds. The availability of each is affected by farm work, by farming systems, and by seasonal change. Many species may at times take crop plants and seeds but these are major food sources only for a few species, notably geese, pigeons and Rooks. The presence of farm stock provides valuable sources of food and water, particularly in winter, when this may significantly influence survival in small passerines. Nearly all farmland species make use of the feeding opportunities arising through farm work. Feeding behaviour varies seasonally and regionally, in line with changes in food availability. The extent to which many passerines use fields as feeding areas varies from place to place, probably depending on the overall pattern of available resources.

5: Modern agriculture: the consequences of specialisation

Moore (1980) identified six major changes in the British countryside that were due to the intensification of agriculture and forestry since the Second World War and that had a major effect on birds. Those affecting agriculture are:

(1) the conversion of broadleaved woodland to arable
(2) the reclamation of heath and scrub
(3) the reduction in the total area (mileage) of hedgerow
(4) improved drainage of large areas of low-lying land
(5) the greatly increased use of chemicals, including insecticides, fungicides and herbicides.

The list undoubtedly indentifies the principal areas of habitat loss in lowland farmland, although the first two are of local significance only, having an impact on agricultural birds only where locally concentrated. Thus loss of woodland to arable nationally involves only a small percentage of agricultural land but in Cambridgeshire, for example, 17 per cent of woodland has been lost since 1945 and in some Scottish counties the figure is as high as 30 per cent (Stubbs, 1980). Similarly, the reclamation of heath and scrub is of minor significance within British agriculture but has caused great conservation concern for its effects on species such as Wheatear which are specialists of these habitats. Upland agriculture does not feature in Moore's list above. Were it included, then increased drainage, reseeding and overgrazing would be included as significant factors.

Moore's list above includes only one aspect of management, namely pesticide use. To these we would add the major changes that have taken place in farm management, even of farmland devoted to the same nominal use. Many of these were not fully developed when Shrubb (1970) reviewed the changes which occurred in farmland bird populations between 1940 and 1967. This review was based on the surveys by Parslow (1968) and suggested that changes in agricultural practices had resulted in general increases in three species, in general declines in seven species, and in local declines in the case of 12 species. Agricultural improvement produced rather more local than general changes, and drainage and other loss of specialised habitat and chemicals were the most frequent cause of decline given. However, this analysis ignored rare species such as Stone Curlew (Morgan, 1986) and also ignores the possibility of simultaneous climatic change. We have since documented various regional changes in bird populations associated with reduced diversity of farm practice (O'Connor & Shrubb, 1986).

Mixed farming systems, combining tillage crops and stock, frequently emerge in our analyses as most favourable to birds. Yet farming has changed dramatically over the last 40–45 years, from an industry whose methods and systems were relatively favourable to wildlife to a specialised and highly technical business heavily biased against the maintenance of the diversity of nature sought in conservation. 'Modern farming is the antithesis of the image projected by the old adage about sowing corn – one for the rook; one for the crow; one to rot and one to grow. Nowadays the farmer sows 'one to grow'; competition within the industry allows little margin for natural wastage.' (Wright, 1980). The high degree of specialisation now evident in modern farming is based in mechanisation and chemical technology. These have combined to remove the problems of weed and disease control, of fertility and of timeliness which led to the earlier development of mixed rotational farming.

Fig. 5.1 illustrates this pattern by comparing the diversity of farming practice in Britain in the 1870s with that of the 1980s. The diversity index used is based on the distribution in each period of cereals, root crops, temporary grass, and sheep and cattle, and shows very clearly the modern trend away from the even distribution of enterprises (and therefore habitats) typical of a mixed farming regime. Instead, we now see large regional blocks dominated by certain crops or enterprises. The distribution of root crops shows the trend most clearly. Forty-seven per cent of the British root crop acreage is now concentrated into the East and East Midland regions of England, reflecting both the increasing importance of major cash crops, such as sugarbeet and potatoes, and the decline of turnips, formerly so necessary in rotation farming as the winter mainstay of the breeding ewe flock. The figure shows that over the last 100 years diversity has fallen everywhere except in the north-west of Scotland. Specialisation has risen not just in East Anglia, where cereal intensification has been recognised as a problem, but also over much of eastern and south-eastern England, in Wales, in the Yorkshire–Lancashire region and

Fig. 5.1. The geographical pattern of diversity of farming practice in Britain in 1873–75 and in 1978–80. The Shannon–Weaver diversity measure was computed over four major categories of farming practice, namely cereals, root crops, leys, and sheep and cattle.

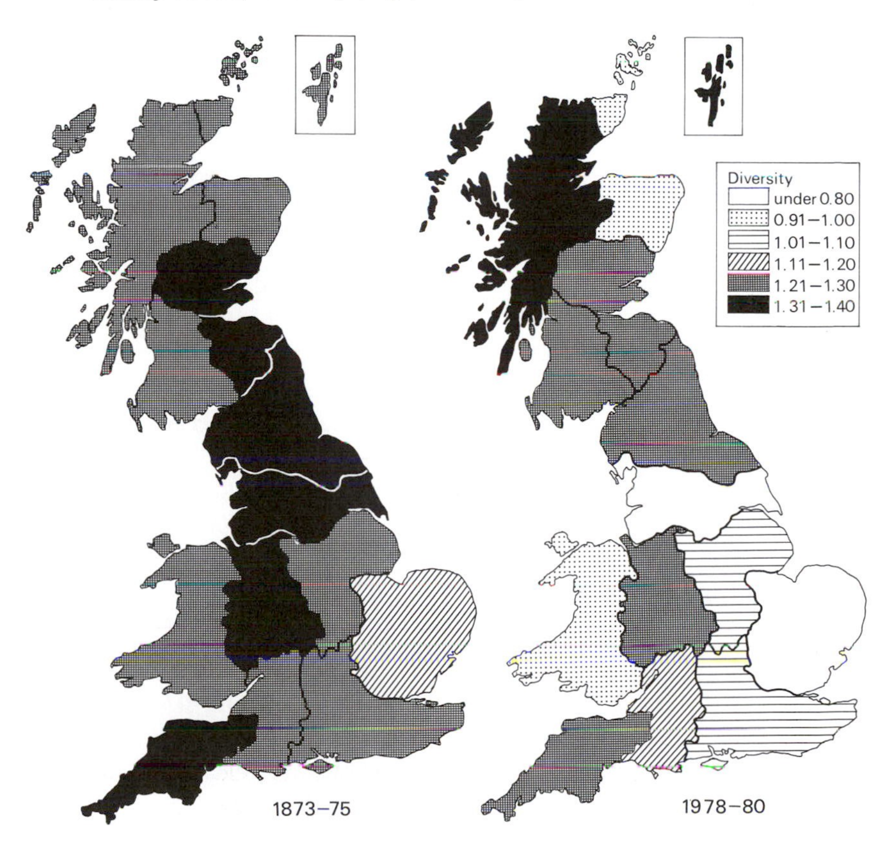

in the north-east of Scotland. Smaller reductions in diversity are shown for the rest of the country. This regional simplification and standardisation of farm practice is, we contend, at the heart of major effects on bird populations. It is the single most important change in farming practice affecting wildlife and is as relevant in grassland farming as it is in arable farming.

The impact of mechanisation

Mechanisation has had two widespread impacts on agriculture, first in replacing men and draught animals by machines in farming operations and, second, through its effects on markets. Labour costs have long been a major item of expenditure for farmers but have been reduced in real terms by recourse to machinery. On Oakhurst Farm, for example, 14 regular staff, supplemented by seasonal casual labour, were employed in the 1940s and 1950s, at an average annual cost of 31 per cent of sales. Such a labour force in 1982 would have cost about 50 per cent of greatly increased sales. In fact, in the 1980s labour cost *c.* 10 per cent of sales. The same point

Table 5.1. *The progress of mechanisation in England and Wales, 1930–81*

	Labour[b] (1000s)	Work horses[c] (1000s)	Tractor mounted sprayers	Combines	Grain driers	Tractors	Root harvesters
1930	630	683				20000	
1940	502	541	1100[d]	1000	190[d]	101500	510[d]
1950	582	289	4188	10048	1309	295261	1318
1960	406	—[e]	51120[f]	47930[g]	16520	416725	11815
1970	276	—[e]	66100[h]	51800[i]	35700[j]	413180	23970[k]
1977	226	—[e]	68130[i]	48390	42730	418700	20990
1981	196	—[e]	74140[l]	47473	30371	420731	24770

[a] MAFF does not gather statistics of all these categories annually.
[b] Labour comprises regular workers.
[c] Horses include breeding mares.
[d] Data for 1942.
[e] Horses were not recorded for the June Census, but there were 21000 in 1965.
[f] Data for 1962.
[g] 30800 of these machines were self-propelled.
[h] Data for 1969.
[i] Data for 1971.
[j] Data for 1968.
[k] Data for 1972.
[l] Data for England for 1978 only.
Source: MAFF statistics.

is also clear from Table 1.3 in the opening chapter, which showed that the purchasing power of a wheat crop in terms of wages declined by *c.* 17 per cent, despite yields increasing nearly three-fold and prices increasing 17-fold. These patterns were general within the industry. There is no doubt that Government capital grants, capital allowances, and the Government support of farm incomes have done much to encourage the process of mechanisation in pursuit of the cheap food policy.

Table 5.1 sets out the progress of mechanisation by measuring the numbers of certain key machines, draught animals and labour at intervals since 1930. In the later years the numbers of some machines have remained relatively stable but their size and power – and therefore capacity for work – have increased. Two simple examples illustrate the trend. In 1967 20 per cent of tractors were of more than 50 h.p. whilst in 1982 the figure was 52 per cent. In 1965 45 per cent of combines had cutter-bars 10 feet or more wide whilst in 1976 the figure was 68 per cent. This trend has been associated with a steep decline in the full-time labour employed. The increasing size and sophistication of farm machinery is well illustrated in Fig. 5.2. Here four tasks – seedbed preparation, spraying, fertiliser incorporation, and planting – are combined in a single pass. The change in the capital structure of farms resulting from mechanisation has done much to promote specialisation in crops and enterprises. On every farm, labour-intensive crops or jobs either have been mechanised or have been eliminated, and the total number of farming enterprises has declined as the capital investment required has increased.

For modern arable farms to be viable with low labour inputs, farm structure must allow the most efficient use of large machines. The impact of such large machinery on farmland practice has been magnified by the growth in the average size of farm holdings in Britain (Table 5.2). Between 1875 and 1979 farm sizes have increased by a factor of 2.0–2.5, thus bringing larger blocks of land within the control of single individuals. The personal attitudes of these individuals to farming then determine the nature of ever-larger blocks of land, and the benefits to wildlife of the diversity of ownership of earlier years are removed. The agricultural support system in fact favours such amalgamation of smaller holdings by its support for ever-increased production: larger farms make better use of larger machinery and in turn generate greater margins from fixed overheads (Sturrock & Cathie, 1980).

Mechanisation may have a direct influence on birds in two areas. First, the increasing size and speed of machines, which tend much more to divorce the operator from his immediate surroundings, may well increase nest and brood losses in ground-nesting species. Second, and probably more important, are the effects of mechanisation in narrowing the complexity and timing of crop rotations.

Mechanisation has had a particular impact on rotations, for these have always reflected the power and technology available to farmers. Rotations serve two purposes – first, to clean the ground and promote fertility, and, second, to spread the work load. Thus the classic 4-course Norfolk rotation not only provided cleaning crops (roots) and compost or manuring crops (clover) but also provided for the most economic use of labour by spreading the farm's work fairly evenly through the year. A major result of mechanisation has been to remove the latter constraint. In cereal farming, for example, the bulk of the crop is now planted in autumn, despite the work load this imposes. Formerly the work used be divided fairly evenly between spring and autumn. The resulting change in the timing of cultivation has also been promoted by the decline in root crops in many

Fig. 5.2. An example of modern cultivation machinery, showing the capacity to perform several operations in the course of one pass over the land. The four operations shown are (from left to right): applying pre-emergent herbicide, seedbed preparation, applying fertiliser, and sowing seed. Reproduced by courtesy of Lely Import Ltd.

Table 5.2. *Changes in the size and number of holdings of crops and grass in Britain between 1875 and 1974*

	England and Wales					Scotland				
	Number of holdings of size					Number of holdings of size				
	0.25–50 acres	50–300 acres	300–1000 acres	Over 1000 acres	Average holding acres	0.25–50 acres	50–300 acres	300–1000 acres	Over 1000 acres	Average holding acres
1875	333630	120164	15633	473	57.0	56311	21701	2658	126	57.0
1935	239167	128633	11618	309	65.7	49147	22822	2366[b]		61.8
1960	210851	120236	13198	651	70.8	38453	20147	2708		70.6
1974	96692	94166	16193	1483	113.4	11484	15738	3377	128	133.4

[a]After 1974 the basis of recording these statistics was changed, making further comparison impossible. By 1979, however, the total number of holdings of crops and grass in England and Wales had fallen to 191 334, averaging 122.7 acres; in Scotland the figures were 29 392 holdings, averaging 140.5 acres.
Source: MAFF June Census statistics.
[b]For 1935 and 1960 the agricultural statistics for Scotland cite figures only for farms above 300 acres

areas and by the increasing concentration on cereals generally. The switch from spring-sown to winter-sown barley, for example, occurred in 39 per cent of the barley area in England and Wales (a total area of 1.65 million ha) by 1980 and just over 50 per cent in 1983. At the same time, wheat has increased to 46 per cent of the cereals area, 98 per cent of it planted in winter (to be compared with the 31 per cent – 87 per cent of it planted in winter – of the early 1960s). In counties primarily concerned with cereal farming, these changes probably affected anything from 16 per cent (as in Wiltshire) to 32 per cent (as in Oxfordshire) of the total land area. Altogether, the MAFF statistics suggest that the percentage of spring cultivation or tillage had declined from *c.* 78 per cent of the tilled land of England and Wales in 1962 to *c.* 36 per cent in 1982. These changes are illustrated in Fig. 5.3.

The seasonal pattern of farm work is something to which every animal inhabiting farmland must adjust, but changes in the periods of tilling the soil have a significant impact on the availability of food supplies to birds, particularly in the breeding season. Spring cultivations are a major source of invertebrate food for many birds when the start of breeding may be limited by the availability of food (Perrins, 1970). Autumn-sown fields are unsuitable feeding areas in spring since the vegetation is by then too tall and dense. Ground-nesting birds such as Lapwing may also prefer spring cultivations for nesting (Klomp, 1953; Galbraith & Furness, 1983).

Another important effect of mechanisation has been to compress the period in the farming year over which food is available to granivorous birds such as Rooks which have depended heavily on cereal harvests. Prior to mechanisation harvesting involved several stages – cutting, stooking, carting, stacking and threshing – which were spread out in time and each of which resulted in some spillage and waste of grain which was subsequently available to birds. Modern combining, in contrast, starts later and yet is completed in a single operation in a shorter time and with less waste than with the older methods. Such efficiency has adversely affected the populations of several species, as discussed for the Rook in Chapter 8.

The mechanisation of transport has profoundly affected farming markets. The impact of major improvements in ocean transport in the free trade era has already been noted but major changes have also taken place within Britain. At the turn of the century virtually all road transport was horse-drawn and even in the early 1930s, when *c.* 5 per cent of road transport remained horse-drawn, farms themselves still owned 1.1 million horses, with 800000 of them in work (Smith, 1949). The demand for hay, straw and oats which this force of draught animals created did much to underpin farming in the recession of the last 25 years of the nineteenth century. The steep decline in the use of draught animals after the 1914–18 war intensified the depression of the 1920s and 1930s. The value of this market is suggested by the Oakhurst farm accounts of the 1940s, when it still accounted for at least 15 per cent of sales (compared with none today). The final disappearance of this market around 1960 partly explains the modern practice of burning straw in the field – the largest traditional market for surplus straw no longer exists. The immediate consequences of burning are to attract birds to consume the invertebrates exposed by the

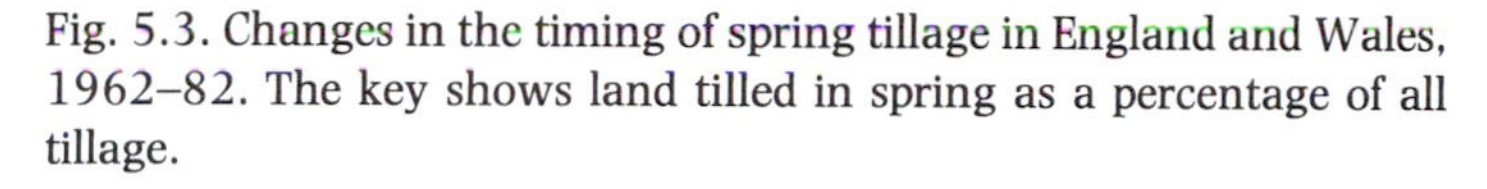

Fig. 5.3. Changes in the timing of spring tillage in England and Wales, 1962–82. The key shows land tilled in spring as a percentage of all tillage.

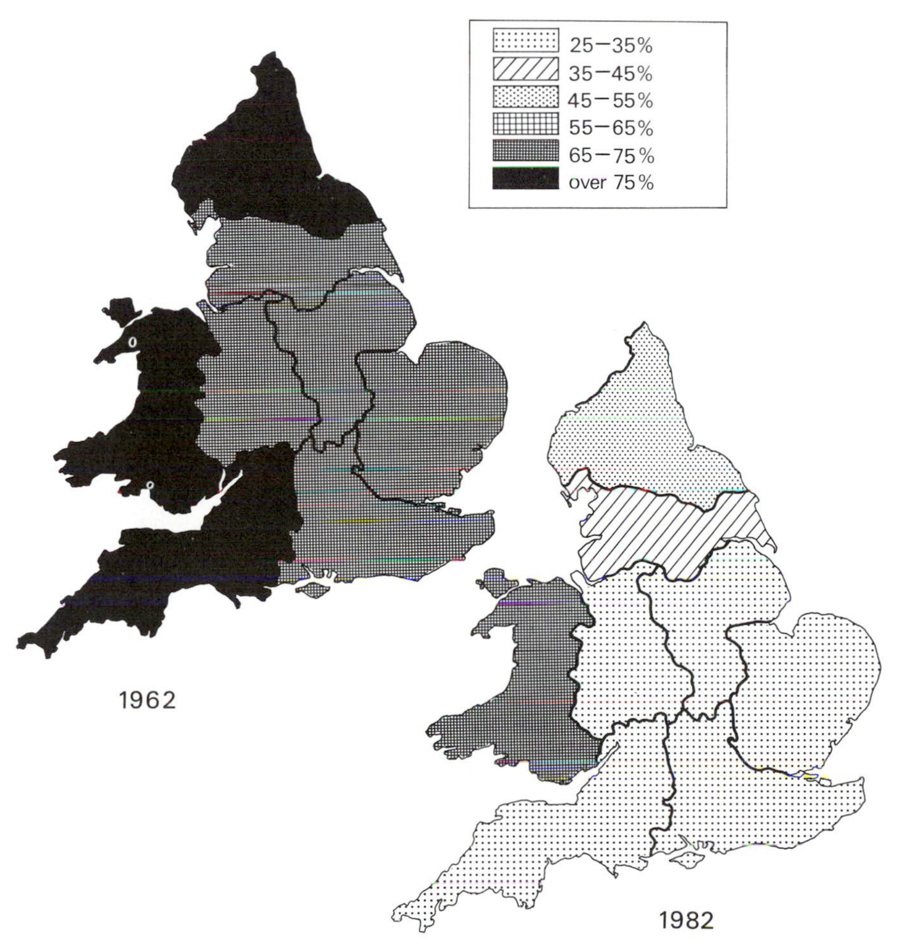

incineration of the straw. In the longer term, however, burning reduces the populations of invertebrates available to birds through the year (Edwards, 1984). The surface-living fauna is particularly badly affected, with various mites, springtails, millipedes and beetles becoming less numerous and taking some time to recover subsequently because of the destruction of the straw habitat. Soil animals are little affected in the short term since the flames are too short-lived to heat the soil to lethal temperatures, but Edwards and his colleagues found an important long-term effect. Deep-burrowing earthworms such as *Lumbricus terrestris* were virtually eliminated from fields burned annually because of the gradual decrease of organic material in the soil. Straw-burning similarly reduces the population of weed seeds in the soil. For example, the flames incinerate as much as 40–80 per cent of black-grass and wild oat seeds present in a field (Chancellor, Fryer & Cussans, 1984) and these figures may be typical of plants with seeds taken by finches. A second and possibly more important effect is to induce earlier germination of many of the remaining seeds: these can then be destroyed by use of a herbicide.

It is difficult to assess the significance of these findings for bird populations. In principle, the increase in autumn drilling in recent years has increased the amount of straw burned and therefore the area of land carrying lower invertebrate and seed populations, in turn possibly supporting fewer and less successful breeding attempts in spring. But, as noted above, autumn sowing reduces the area of spring cultivations, a major source of food at the start of the breeding season. Hence both the abundance and the availability of food are affected by the trend to autumn sowing and the immediate cause of the declines in ground-feeding birds

Table 5.3. *Changes in the distribution of sheep in arable and upland regions in England and Wales 1873–75 and 1981–83*

	Arable regions % of total flock			Upland regions % of total flock	
MAFF region	1873–75	1981–83	MAFF region	1873–75	1981–83
East	12.8	1.4	West Midlands	8.6	10.1
South-east	14.3	6.9	South-west	17.1	13.9
East Midlands	13.1	5.7	North	12.3	22.5
Yorkshire & Lancashire	7.2	4.2	Wales	14.3	35.3
Total	47.4	18.2	Total	52.3	81.8

(p. 162) may be due to either. *A priori*, the changes in spring feeding are more likely to have immediate effects and to account for the speed of the declines recorded.

Improved transport has also greatly assisted the modern trend to specialise on farms, since the successful concentration of production needs good and rapid bulk transport to distribute perishables. This was particularly an essential prerequisite for the vast expansion of the liquid milk market during the past 100 years. Dairying was fully established as a major branch of farming only at the beginning of the twentieth century (Trow-Smith, 1951) and its steady development since, strongly boosted by the poor economics of corn growing in the 1920s and 1930s, has led to marked changes in the farmland scene. Most noticeable have been: the decline of sheep in the arable regions (the present upland concentration is shown by Table 5.3 to be a modern phenomenon), the consistent increase in cattle numbers everywhere throughout the last 120 years, and the dominance of barley in the cereal acreage as the basis of cattle feeds. Silage, cattle and barley have replaced hay, roots, sheep and oats in the arable areas.

Chemicals have combined with mechanisation to remove the constraints that historically led to the use of mixed farming techniques. Mechanisation eliminated the need to spread the demand for labour by use of rotations. However, the primary function of rotation has always been clinical, to clean the farm of weeds and the risks of disease. Since 1945 these functions have increasingly been carried out by chemical means, using an ever-widening and sophisticated range of seed dressings, fungicides and herbicides. In the same way the problem of feeding crops has been resolved by the use of chemical fertilisers, which are formulated for particular and precise applications. The harnessing of chemical technology is the one new element in modern agriculture and is discussed in detail in Chapter 9.

Changes in permanent grassland

Post-war agricultural policies were successful in reversing the effects of the agricultural depression. By 1983 the arable area had been restored almost to the 1875 level. This was achieved very largely in the 1940s and 1950s by paying grants to farmers to plough permanent grassland. These grants amounted to £7.50 per acre (about £112 at 1985 prices) and when combined with cereal deficiency payments were a sufficient inducement to encourage the necessary investment in machinery to convert 4–5 million acres (1.6–2.0 million ha) of permanent grassland back to arable by the early 1960s (Table 1.1 above). Fig. 5.4 shows the regional variation since 1875 in the arable/grassland balance. Indeed, it would be possible

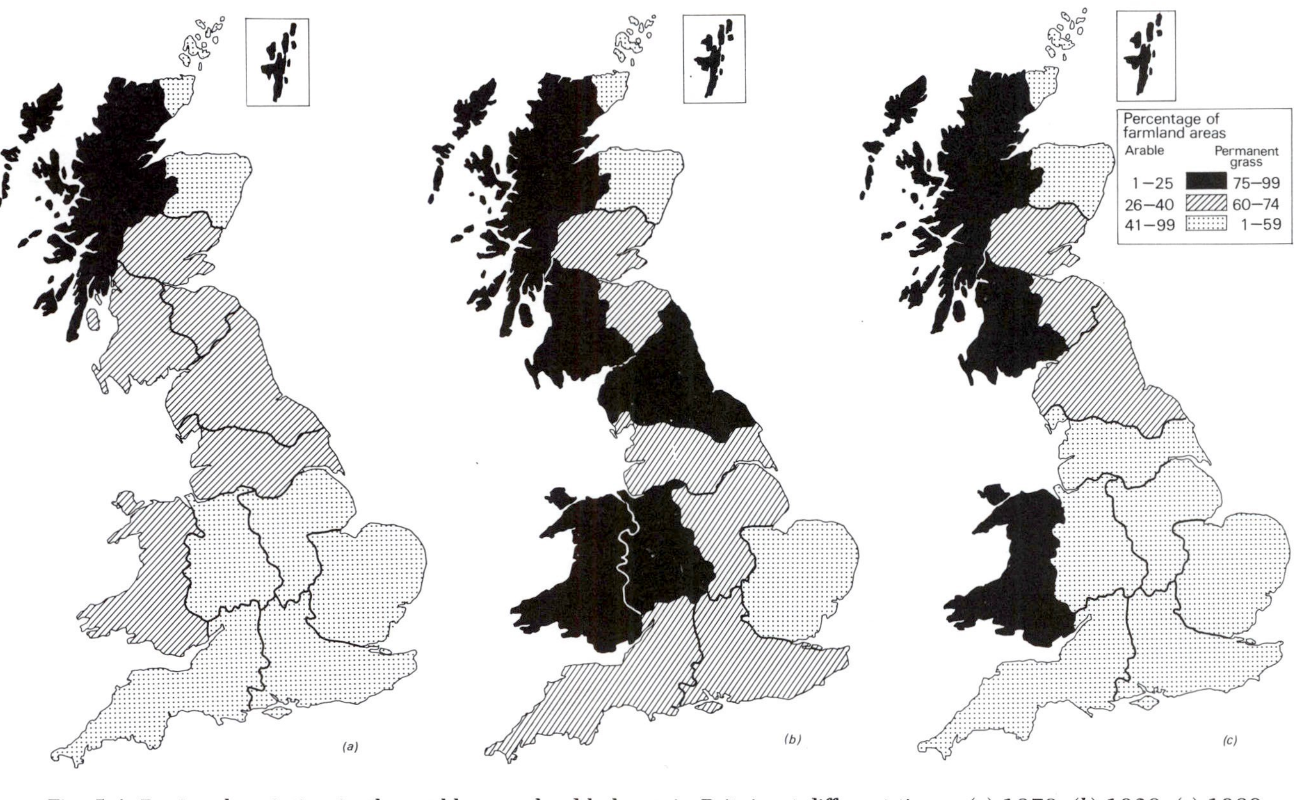

Fig. 5.4. Regional variation in the arable–grassland balance in Britain at different times: (*a*) 1870, (*b*) 1930, (*c*) 1980.

to demonstrate such variations in the relative acreages of grass as far back as the twelfth century. Variations in the arable area of Scotland were much less marked, very largely because Scotland has comparatively little good cropland. Just as the increase of grassland and the decline in its quality in the depression was most marked in the arable lowlands of the south and east, so the decline of grassland habitats subsequently has been most marked there. But the recent changes have also led to a new shift in the distribution of arable land which has in turn caused the loss of important habitats, notably chalk grassland and sheepwalk.

These changes had major consequences for birds. Fig. 5.5 shows the results of a Danish study that examined how various species were affected by the transition to cereal production in one area of Jutland. Over a seven-year period Yellow Wagtail populations decreased by over 80 per cent, whilst at the other extreme Whinchats decreased by only 11 per cent. The figure shows that these changes were strikingly linked with the location of territories in relation to grassland and corn fields, so that birds that normally located their territories in grassland suffered most. In Britain two related species, Wheatear and Whinchat, provide good examples of the impact of increased arable farming on birds in sheepwalk and meadows. In Sussex and Wiltshire a drastic decline of Wheatears followed the war-time and post-war ploughing of the Downs, confining the remaining pairs to the steeper slopes. In Sussex the spread of the coastal towns aggravated

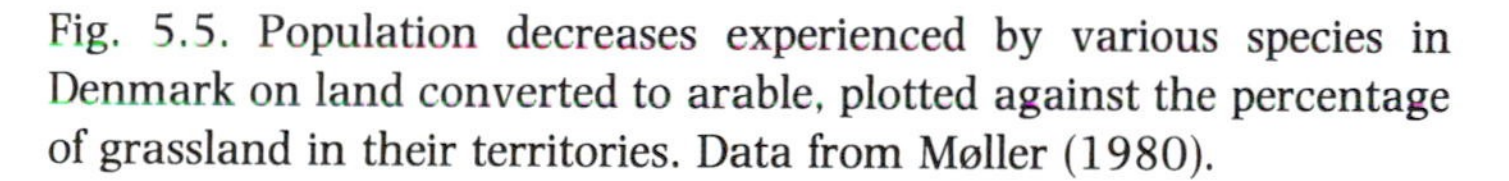

Fig. 5.5. Population decreases experienced by various species in Denmark on land converted to arable, plotted against the percentage of grassland in their territories. Data from Møller (1980).

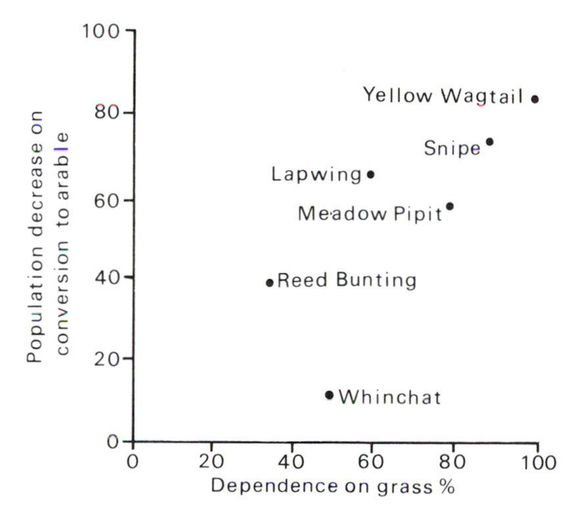

the process. In both counties further declines followed the myxomatosis epidemic of 1954 which left fewer rabbits grazing the vegetation to the short sward favoured by these species; fewer burrows meant that nesting holes also became scarcer (Shrubb, 1979; Buxton, 1981). Similar declines were noted in the Cotswolds (Swaine, 1982) and further north wherever hilly pastures with short turf and rabbit holes were ploughed in response to war-time need and post-war grant. In the Midlands, however, the impact of such change was less marked, possibly because the Wheatear had already been reduced in numbers by enclosure and cultivation in the latter part of the nineteenth century, as in Derbyshire (Frost, 1978). Indeed, Alexander & Lack (1944) recorded a long-term decline of Wheatears in England through the previous century and the most recent changes thus represent an acceleration of an existing trend. Nevertheless, in Devon there has been little recent change in the bird's distribution (Sitters, 1974) and here the MAFF statistics show a change in the quality rather than in the area of grass.

Whinchats feed on insects from field-layer vegetation, especially from umbellifers, and often nest in mowing grass. They therefore found an abundance of attractive habitat during the 1920s and 1930s as neglected pastures were colonised by small hawthorns. In Nottinghamshire, for example, the Whinchat spread until this habitat was destroyed by post-war farming (Dobbs, 1975). This process has been fairly general in lowland England (Parslow, 1973) and has been specifically noted, for example, in the West Midlands (Harrison *et al.* 1982), in Derbyshire (Frost, 1978), in Gloucestershire (Swaine, 1982), and in Kent (Taylor, Davenport & Flegg, 1984). As with the Wheatear, many alternative areas of suitable breeding habitat – tussocky grass with thinly scattered tall weeds or bushes – vanished under a dense cover of scrub following the widespread eradication of rabbits by myxomatosis after 1954. Elsewhere, the increase in herbicide use and the switch from haymaking to ensilage have adversely affected the species in agricultural habitats, whilst in the wider countryside the modern management of road verges, a much-used habitat in central England, has had further impact (Harrison *et al.*, 1982; Swaine, 1982).

Changes in rough grazings

In general terms rough grazings are areas of grassland having little or no management input. Although the term lacks official definition, many of the most important grassland habitats for birds – marshes and wet meadows in the lowlands, moorland in the uplands – are probably classified as rough grazings in the MAFF statistics. Drainage grants and liming grants (which are of particular relevance to upland farms) have encouraged the reclamation of an area which the official statistics suggest is not less than 500000 acres (202 500 ha), this being converted from rough grazings to crops and pasture, largely the latter. In grassland this process almost invariably means reseeding. MAFF calculated in 1968 that, in addition to these changes, 20000 acres (8100 ha) of rough grassland a year, mostly in upland areas, have been converted to forestry since 1946 (HMSO, 1968). The rate is unlikely to have varied greatly since. Fig. 5.6 illustrates the variation in the area of rough grazings since the late nineteenth century when such statistics were first gathered. The figures for rough grazings, particularly for the early years, must be treated with some reserve since the assessment of this category is a subjective one made by the farmer. As a declaration of intent, however, changes in this category are probably the truest indicator of business confidence among farmers. Moreover, there is little reason to doubt that the area of rough grazing in 1930–32 was historically at a very high level.

The regional variations in reclamation indicated by Fig. 5.6 are of considerable interest. In Scotland the overall area of rough grazing is still similar to that of the late nineteenth century and in the Highlands the area has actually increased. This is probably a social change as people have chosen to move to a less demanding environment. In England and Wales major changes in land use are involved. In the predominantly arable areas of the east and south a combination of natural fertility, favourable climate, economic incentive and powerful drainage machinery has resulted in a significant percentage of wet grassland being reclaimed for arable. Here the total numbers of sheep and cattle, expressed as grazing units, have increased by only 3 per cent since 1930–32. In the west reclamation has been primarily for pasture improvement and stock units have increased by an average of 73 per cent. Overall it is in the rough grassland area that agricultural pressure on valuable habitat is greatest and the loss of prime land in the Depression most damaging to the conservation interest.

In the English lowlands rough grazings are very often wetlands and these areas are among the first to react to economic recession in farming because of high maintenance costs. They are steadily reclaimed during economic prosperity because of their high natural fertility. Redshank and Snipe are the two breeding species most severely affected by these changes. Redshank need a suitable feeding area near the nest, which is usually in a tuft of long grass, and these requirements limit the species inland to rushy fields and waterside meadows, and to similar habitat at reservoirs and gravel pits. Coastal grazing marshes are a particularly important habitat in maritime

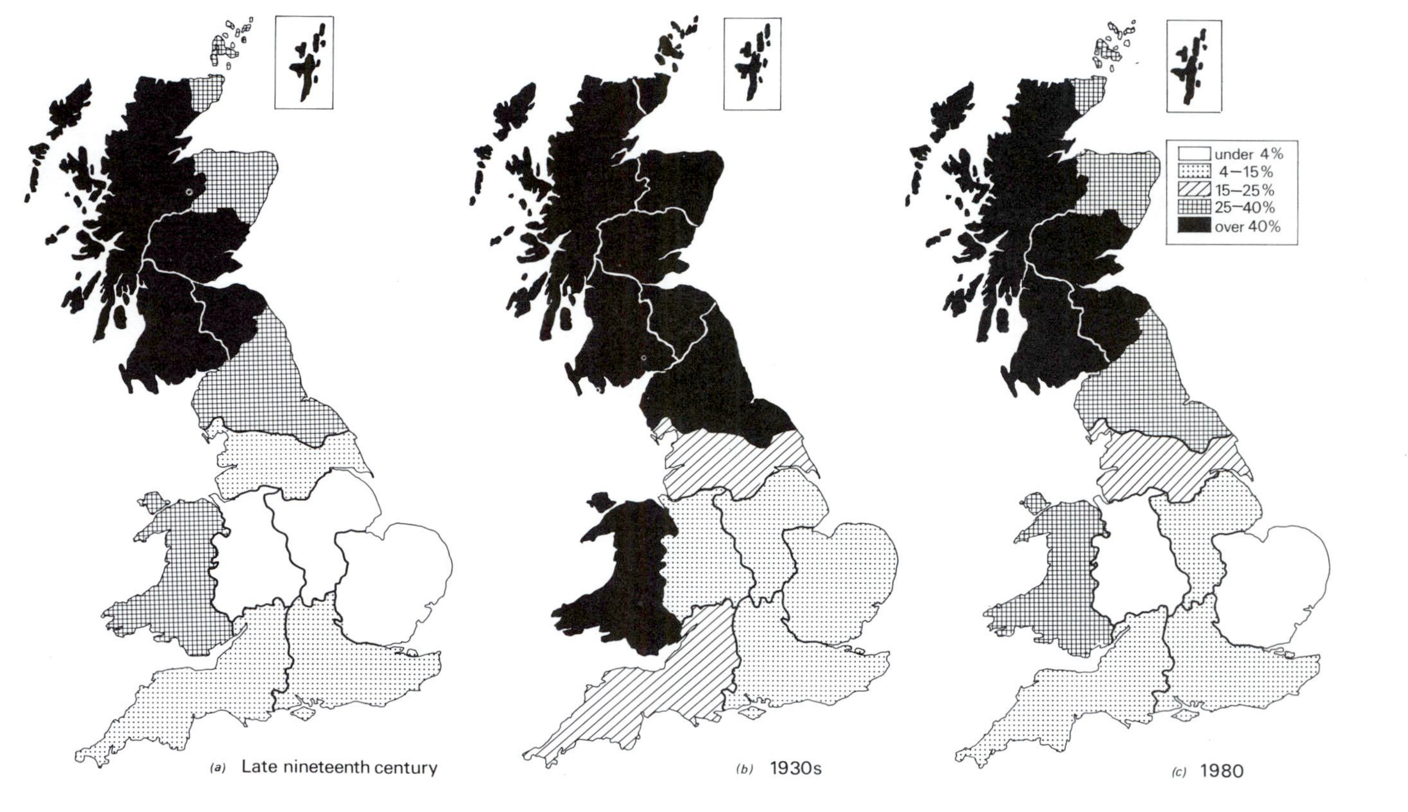

Fig. 5.6. Regional variation over the last 100 years in the proportion of rough grazings on agricultural land in Britain: (*a*) late nineteenth century, (*b*) 1930s, (*c*) 1980s.

counties and may support very high concentrations locally, particularly in East Anglia and in north Lancashire and Cumbria (Smith, 1983). Redshank declined greatly in the face of early nineteenth century drainage but by the latter half of the century the species was spreading inland again, reaching the Midlands by around 1885. This coincided with the sharpest phase of the agricultural depression of 1875–1914, which was partly triggered by a series of appallingly wet years and harvests in the late 1870s. Between 1885 and 1895 permanent grass in arable districts of England rose by nearly one million acres (0.405 million ha) (Orwin & Whetham, 1964) and the total area of rough grazing rose steadily from 1892 to the 1939–45 war, when it fell very steeply. Redshank numbers varied in parallel, as shown for Oxfordshire in Fig. 5.7. The species is thus tightly coupled to the intensity of grassland management and to the subsequent habitat changes. This pattern is widely documented in county avifaunas and both Hickling (1978) and Shrubb (1979) specifically noted that drainage and agricultural improvement of permanent pasture rather than the conversion of pasture to arable were responsible for post-war losses in Leicestershire and Sussex. Pasture improvement is usually marked by steep increases in stocking rates and the resulting tighter grazing of the sward destroys the tussocks used as nest sites by Redshank. A national survey

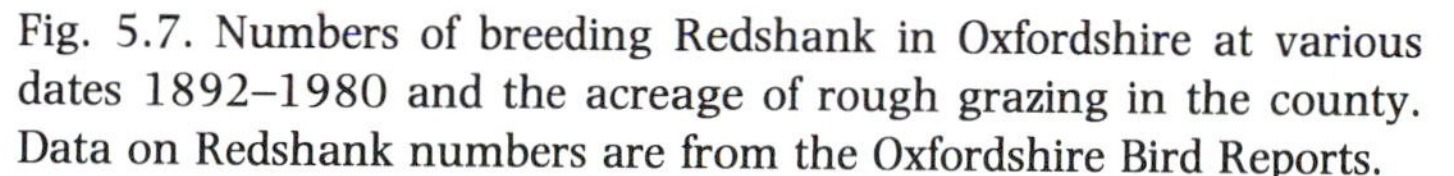

Fig. 5.7. Numbers of breeding Redshank in Oxfordshire at various dates 1892–1980 and the acreage of rough grazing in the county. Data on Redshank numbers are from the Oxfordshire Bird Reports.

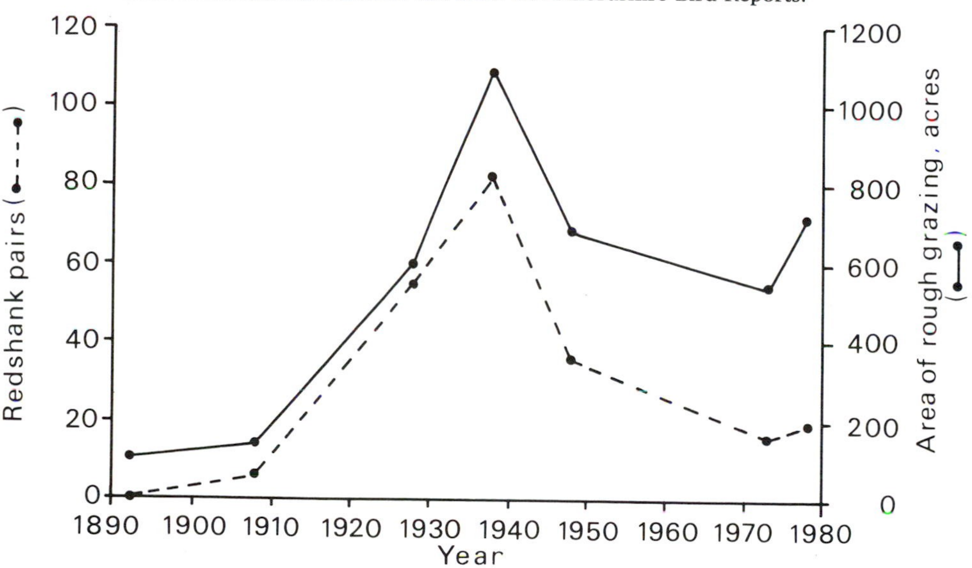

of its distribution on lowland wet meadows (Smith, 1983) shows that its range has once more contracted towards the east coast. Only a third of the sites visited held Redshank and 72 per cent of pairs were concentrated in the 400 coastal sites surveyed compared to 38 per cent on 882 inland sites. These figures exclude those birds breeding on saltmarsh, a habitat in widespread use and known to support high breeding densities (Greenhalgh, 1971, 1973).

Snipe numbers are not as closely correlated with the acreage of rough grassland as are Redshank numbers since Snipe will use only the wetter patches. On the Yare marshes in Norfolk, for example, Snipe were found only on damp peaty margins and not at all on the main grazing marshes (Murfitt & Weaver, 1983). As seen for Redshank above, local avifaunas

Fig. 5.8. The distribution and numbers of drumming Snipe found on grassland sites in each 10-km square during the survey of Smith (1983). The key relates to numbers of drumming individuals per square.

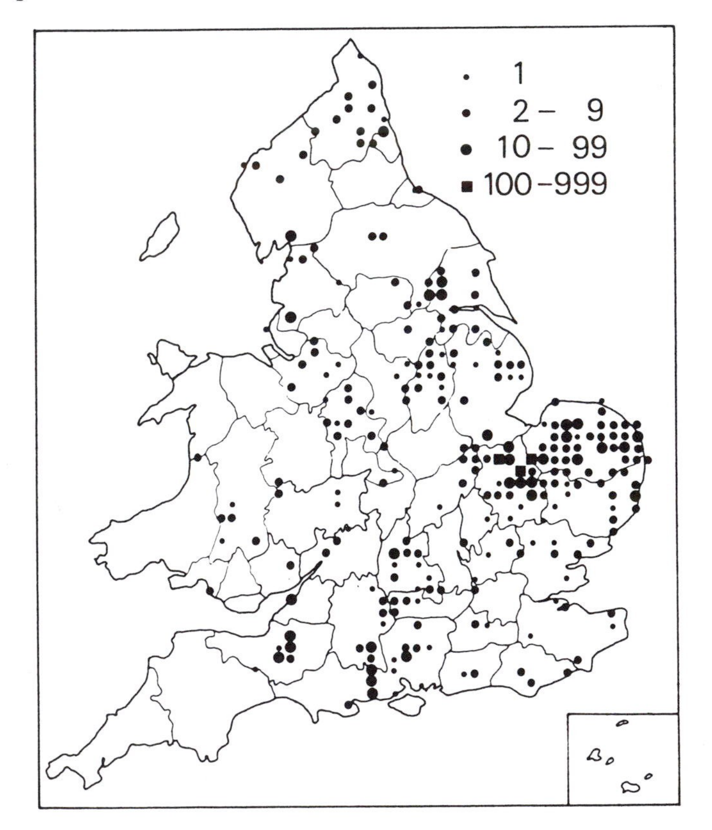

similarly show that Snipe numbers have increased and contracted with changes in farming prosperity (e.g. Parslow, 1973; Dobbs, 1975; Frost, 1978; Shrubb, 1979). The national survey in 1982 showed that very few Snipe – a total of only 1979 drumming birds was found – now remain in lowland wet meadows in England. These are concentrated into a few major sites: no less than 48 per cent of all pairs surveyed were on just five sites – the Ouse Washes, Nene Washes, North Kent marshes, Derwent Ings and the Somerset Levels – and 25 per cent were on the first-named alone! (Indeed, these five sites held 29 per cent of all waders found on grassland in that survey!) Fig. 5.8 shows the resulting strong easterly distribution of this species. Smith (1983) suggests that this distribution is due to western valleys being generally dry, with only occasional flash floods and rather few areas of peat. In contrast, rivers in the east usually have high water levels, in turn making for high watertables.

In the uplands much of the loss of rough grazings has been to forestry, the only viable alternative to stock farming. Newton (1983) has reviewed

Fig. 5.9. Population densities of selected species in relation to age and growth stages of Sitka spruce plantations during afforestation of sheepwalk in southern Scotland. Data from Moss *et al.* (1979).

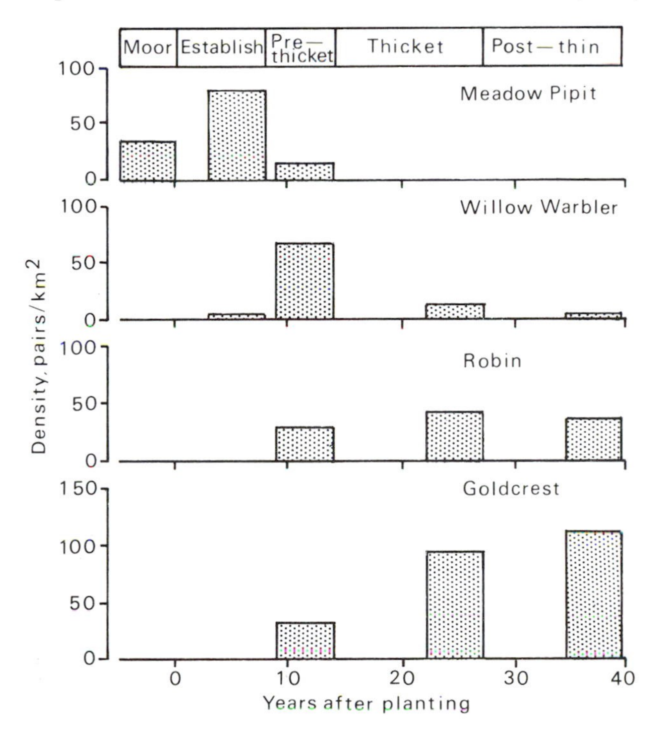

the effects on birds of conversion of sheepwalk to plantation. On unplanted grassland grazed by sheep, Skylark and Meadow Pipit dominate at combined densities of 70–130 pairs per km^2, levels rather higher than the 40–70 pairs per km^2 of heather moor (Moss, Taylor & Easterbee 1979). As a result of the planting and subsequent growth of conifers Skylark numbers decrease, Meadow Pipits at first increase, then decrease, and various other passerines appear (Fig. 5.9) In this way the number and variety of songbirds increase but no one species is present at all stages from open land to mature forest. Ravens disappear about the time of planting, largely due to the reduction in sheep carrion (Marquiss, Newton & Ratcliffe, 1978). Waders such as Curlew are lost before the pre-thicket stage (trees 12–14 years old) whilst Merlins disappear by the time the trees reach thicket stage (23–27 years). Conversely, vole predators are particularly numerous in the early stages of the plantations, for the removal of sheep allows the grass to grow and leads to densities of *Microtus agrestis* as much as 100 times higher than on sheepwalk. Hence Short-eared Owl, Barn Owl, Long-eared Owl and Kestrel may all increase, though Barn Owls are also particularly dependent on the presence of abandoned farm buildings and cottages. Once the grass is shaded out by the growing trees, vole numbers and the associated birds decline.

Where forestry replaces sheepwalk only locally, additional habitat diversity is introduced, and total bird density may increase. This occurs not only because new species live within the plantations but also because a number of raptors benefit from the provision of forest where they can breed alongside open ground over which they can hunt. Examples include the Hen Harrier, Buzzard and (locally) Goshawk. As planting continues, however, these species are lost as the proportion of open ground decreases. Consequently the major impact of afforestation on birds of conservation interest has been where blanket forestry has replaced sheep rearing, as in Galloway and Northumbria (Nature Conservancy Council, 1977; Newton, 1983).

Changes in temporary grass

The third component of grassland farming is temporary grass or ley. Such grassland reached a peak of 2.4–2.8 million acres (0.97–1.13 million ha) in the 1950s and the early 1960s, when 51 per cent of the area was concentrated in the West Midlands, South-West and Welsh regions. It has declined since as stock farmers have dispensed with arable production altogether, and instead buy in whatever cereal-based foodstuffs they require. The most important loss here is the decline of one-year clover ley.

Until the 1930s much of the area of temporary grass was probably one-year leys, largely of clover, at least in the east and south. Such leys had declined to no more than 7.9 per cent of the total temporary grassland area by 1974 and to less than one per cent by 1982. Thus the bulk of the area is in longer period leys and an important percentage may, in fact, never be any crop other than grass regularly reseeded in rotation. But the farming distinction between ley and pasture is worth retaining as there are significant differences between them in botany and in management. Grass leys are likely to receive much higher dressings of fertiliser and are twice as likely to be mown than is pasture (48 per cent of the total area compared to 23 per cent). A higher percentage of leys than pasture may also be mown for ensilage (see Chapter 8).

Changes in the availability of grassland have been shown to affect the numbers of Jackdaws and Rooks shot on East Anglian estates (Tapper, 1981). The two species feed extensively on invertebrates on short grassland during the summer and Fig. 5.10 shows how Rook and Jackdaw numbers have changed in East Anglia in relation to the availability of grass leys. The decline in ley grass preceded the decline in numbers of both species,

Fig. 5.10. Changes in the mean numbers of Rooks and Jackdaws recorded from East Anglia in the National Game Census. Shaded areas shows the maximum extent of losses which could have been due to Dutch elm disease. Dotted line indicates the area of ley grass in the MAFF Eastern Region. Redrawn from Tapper (1981).

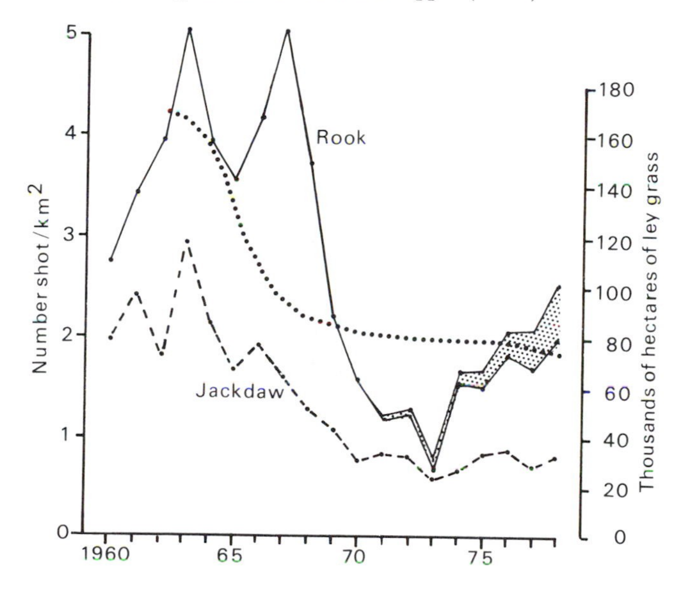

but once the acreage of leys stabilised, so too did Jackdaw numbers, whilst Rook numbers actually increased. An alternative explanation – that Rook were affected by the spread of Dutch elm disease, since up to 60 per cent of Rook nests used to be in elms – is untenable. Data on the spread of the disease showed that the loss of elms occurred too late to account for the decline, though it might have slowed the rate of recovery (Fig. 5.10).

Both Potts (1970) and Shrubb (1970) comment on the value of undersowing cereals with short-term clover or grass leys in rotation. In such farming systems clover was the most widely used ley and its decline may therefore be of particular consequence for certain birds. For example, Potts found that sawflies overwinter most successfully in the leys established under cereals and then migrate into nearby cereal fields. Thus the number of sawfly larvae, an important food for Grey Partridge chicks and buntings, is increased by the presence of such leys in the rotation. For other birds, e.g. Jackdaws and Skylarks, clover weevils are an important food (G. R. Potts, personal communication). Stubbles undersown with clover are also particularly favoured nesting sites for Corn Buntings, as well as for Reed Buntings when they breed in fields. Such stubbles are also heavily used in winter by seed-eating birds. Green (1978) found that clover suppresses dicotyledonous weeds less than does grass, leading to larger numbers of weeds seeding. This advantage is lost, however, if the cereal crop under which the ley is established is sprayed with herbicide.

The increase in cereal acreage

Losses of grassland have been matched by increases in the area of cereal production. In general cereals have moved out of the South-West and Welsh regions and into the East Midlands and Yorkshire–Lancashire regions and the distribution of the individual crops has also changed. Thus wheat is significantly less widely distributed today than in the 1870s, probably as a result of the modern concentration of flour milling into large units served by rapid bulk transport. Oats have declined sharply, from 1.66 million acres (0.67 million ha) in England and Wales and 1 million acres (0.4 million ha) in Scotland in 1875 to 0.22 million acres (90000 ha) and 0.10 million acres (40000 ha) respectively in 1980, reflecting their declining favour in feeding stock. Although overtaken by wheat in southern England, barley is now the most widespread cereal and lends itself to continuous cropping. This and modern farm economics have encouraged the modern phenomenon of the specialised cereal farm and such farms now account for about 20 per cent of cereal production. Such changes in the distribution of individual cereals are probably not critical in themselves but they have an important bearing on the seasonal pattern of tillage, particularly in the barley acreage, which is discussed below.

Intensive cereal production is generally unfavourable to birds, as can be seen from the results of a BTO survey of the breeding birds of Manydown Farm, a modern cereal farm in Hampshire (Fuller, 1984). Production here was modified by a major commitment to habitat management in the interests of gamebirds and this served to highlight the effects of intensive cereal cultivation on small songbirds. In total the 562 ha census area supported about 1200 songbird pairs but of these no less than 53 per cent were located in various woods (mostly oak with hazel coppice understorey) that covered only 11 per cent of the area (equivalent to 969 pairs per km²). Gardens and habitations were also disproportionately populated, holding 18 per cent of the songbirds in just 4.4 per cent of the area (equivalent to 867 pairs per km²). In contrast, the cereal fields and the short and treeless hedgerows surrounding them held only 26 per cent of the songbirds on 71 per cent of the land area, the balance being accounted for by grass and small acreages of other crops. Breeding density on the cereal crop areas was only 21 pairs per km², mostly of Skylarks and a very few Corn Buntings. These figures testify to the poverty of the cereal environment for songbirds. The farm studied by Fuller (1984) was unusual in that it maintained considerable areas of cover in the interests of partridges and these areas then provided both food and breeding sites for small passerines. In the absence of these areas only the rather poor hedges would have been available.

Very few birds made any noticeable use of the crops for feeding. Of the 22 woodland species present, only the thrush species were seen foraging (and then rarely) on the rotovated margins of the field – wood interface. All the birds nesting in the hedges fed in crops along the hedge, however, though most of their food came either from within the hedges or from adjacent gardens or woodland. A strong seasonal effect was present, with several species using the crops more after mid-June, presumably in response to a seasonal increase in invertebrate populations on the crops. Fuller suggested that the woodland species could get adequate food within the woodland and so did not need to use the poorer feeding areas of the fields. Hedgerow territory-holders, however, had to use the fields to some extent because they could not get enough food elsewhere. In a study on arable farmland in Cambridgeshire and its adjacent counties, Arnold (1983) similarly concluded that thrushes and tits fed in woodland wherever they could (especially in winter) and used the adjacent farmland only where woodland was scarce.

Summary

The result of the post-war revolution in agriculture has been a great increase in specialisation in farming. This loss of diversity of farming practice is the single most important factor affecting wildlife on the farm. The most significant effects may be in grassland, where specialised birds have suffered from habitat loss and from widespread management changes. The direct impact of mechanisation is rather limited but machines have been a crucial factor in narrowing the range of rotations practised. Much of the grassland lost has been converted to specialised cereal farming which is a very poor environment for birds.

6: Farm structure and bird habitats

In many ways farmland in Britain can be considered as a matrix of habitat elements (Fig. 6.1). The skeleton consists of hedges, small woods, patches of scrub, waterways, roads and tracks, and farmsteads, all of which are relatively permanent features. Fields are also permanent in this sense and therefore form part of the farm skeleton but their use is determined by the choice of cropping and of techniques that together constitute farm management. In this way each holding has a basic farm structure and a more rapidly changing management regime, each of which affects which species of birds may use the farm. Farm structure can of course be changed, by removing woodland or scrub, by amalgamating fields, by removing hedges, by draining ponds, and so on, but such changes are by their nature infrequent within each individual farm. In this chapter we consider how farm structure influences the local bird communities, with discussion of the influence of management deferred to Chapter 8.

Surprisingly little is known about how habitat features of agricultural land may determine or limit farmland birdlife. On general natural history

Fig. 6.1. Farm habitats in 1971 in an area of mixed farmland at Hambrook in West Sussex. The pale fields are spring-tilled land (cereals and horticultural crops), the dark fields with lines are winter cereals, and those without lines are grass. The same area today is dominated by winter cereals. Habitat features identified are: 1, old hedge lines, removed in the early 1960s; 2, grass marsh with stream; 3, tidal creeks; 4, winter cereals in fields long reclaimed from tidal creeks; the earlier network of runnels is still visible; 5, an isolated farm pond; 6, lines of trees without underlying hedges; 7, some grassland fields;

grounds, conservation advice has focused on the retention of hedges (see below), on the maintenance of the continuity of a hedgerow network to facilitate bird movement out of woodland 'reservoirs', on compensation of essential hedge removals by tree planting in field corners, and on the avoidance of unnecessary destruction of small woodland and wet areas. But as Murton & Westwood (1974) pointed out, much of the conservation advice on offer is little more than folklore and is susceptible to counter-example. Thus even one of the most widely accepted principles – that the retention of hedgerows is essential to the maintenance of local bird numbers – is challenged by Murton & Westwood's finding that the removal between 1960 and 1971 of two-thirds of the hedgerow on a Cambridgeshire farm was accompanied by a net increase in the number of breeding species. We review here, therefore, some quantitative studies of habitat dependencies among farmland birds.

Table 6.1 summarises the results of our invitation to long-term CBC participants to contribute their experiences as to how changes in farm structure have affected the bird communities they record on their census plots. Such a summary of anecdotal evidence is necessarily subjective but in the absence of any better data is still informative. As case histories, the reports reveal the ways in which modern farm practice is simplifying habitat structure. The table shows the very wide spectrum of structural changes that can affect birds, with 15 distinct factors listed. Dutch elm disease and hedgerow loss are the two most frequently cited structural changes, though both have of course received much publicity. The other most frequently quoted factors related to aquatic habitats – the creation or loss of ponds, clearing riverside vegetation, and river management. Such habitats often have a disproportionate effect on the bird interest of a farm. The loss of woodland and the clearance of scrub and ditches were also frequently remarked but the former was in part offset by tree planting. The loss of farmland to various forms of urban development (including road construction) was generally detrimental but in some cases created undisturbed areas of wasteland where weeds flourished for finches. Excluding

8, small woodlands; 9, orchards; 10, scrub; 11, an extensive area of farmstead, cottages, gardens and trees; 12, an isolated set of farm buildings with a shelter belt; 13, close-trimmed hedges without trees; 14, tall well-structured hedges; 15, trimmed hedge with standard trees; 16, a relict hedge; 17, soft fruit nursery, probably mainly strawberries.

the natural disaster of Dutch elm disease, 78 (86 per cent) of the instances reported referred to habitat loss.

Information on the habitat requirements of farmland birds can be obtained from four sources. The first is the study of the distribution of breeding birds within farmland study plots, perhaps best exemplified by the work of Williamson and his colleagues (Barber, 1970; Williamson, 1971). These case studies have mapped the distribution of birds breeding on the farm concerned and have analysed the extent to which territories coincide with features of the landscape. A second and more informative type of study comes from the repetition of such distributional studies before and after modification of the habitat has taken place. Changes in the positioning of birds' territories are likely to reflect the extent to which essential and non-essential elements of the birds' requirements have been affected by the habitat alterations. Examples of this approach include the studies of the impact of hedgerow removal on birds by Murton & Westwood (1974) and by Bull, Mead & Williamson (1976), and Osborne's (1982a,b; 1983) investigation of the effects of Dutch elm disease on bird populations. A third approach, not yet widely used, has been to correlate the location of bird

Table 6.1. *The incidence of structural changes on CBC farmland plots*

Increased urbanisation[a]	7 (2)[b]
Destruction of old buildings	3
Ponds	10 (4)[c]
River management	9
Ditch clearance	9
Riverside vegetation cleared[d]	9
Scrub clearance	8
Hedge loss	12
Dutch elm disease	14
Woodland loss[e]	8
Tree planting[f]	8 (7)
Other[g]	8

[a] Includes road construction.
[b] Two cases of birds benefiting from roadside vegetation along new roads.
[c] Four cases of pond creation.
[d] Includes one case of reedbed destruction.
[e] Includes clear-felling, destruction of orchards, and spinney removal.
[f] Includes one case of 'afforestation' reducing bird populations.
[g] Includes four cases of increased flooding of land, two of general tidying farmland, and one each of heath destruction and of 'loss of trees'.
Figures in parentheses indicate cases favourable to birds.
Data are from a survey of CBC participants' experience (see text).

registrations within a farm with the habitat elements there, as attempted by Osborne (1982a). This is in effect a statistical extension of the distributional studies already mentioned. Finally, analysis of the relationships between the densities of birds present on each of a large number of farmland plots and quantitative estimates of the habitat on each are possible, either with a national data set such as that of the CBC (e.g. Williamson, 1967; Morgan & O'Connor, 1980) or with a series of smaller, locally replicated, census plots (e.g. Arnold, 1983). All four approaches have some features in common but each has particular advantages and disadvantages. The main problem with all of them is that observed changes and/or correlations may always turn out to be spurious and due to some other unmeasured factor altering in phase with the change measured. Nevertheless, we are beginning to see the results of such a variety of habitat studies on British farmland that it is unlikely that results consistently obtained in all of them are incorrect. However, we do still lack studies of how birds use farmland in winter, notable exceptions being the studies of Murton (1971), Arnold (1983), and that of Moles (1975) in Ireland.

Farmland as a bird habitat essentially comprises a recurrent pattern of crops or grass in fields, of field boundaries, and of farmsteads and cottages with gardens and trees. Through this pattern are fragments of other habitats, such as woodland, marsh, streams or scrub, thus producing the varied landscape of the traditional English countryside. Some fragments are remnants of much older habitats but others are of recent origin, as where unploughable slopes on the southern chalk are reverting to scrub because fewer sheep and rabbits control it by grazing. In addition, the fragmentation varies regionally. Both the diversity of the bird community in the area and the absolute numbers of birds present are influenced by the availability of habitat on a farm and both contribute to the ornithological interest of the site. However, although conservation interest is often placed on promoting the maximum variety of species possible, the abundance of individual species deserves consideration too. Consider how strikingly different would be a lowland farm on which Blackbirds or Skylarks (normally the commonest species present) were as rare as Lesser Whitethroats! The sharp drop in the relative abundance of breeding Whitethroats, brought about in 1968 by the onset of severe drought in the winter quarters in the Sahel, provides one real example of the impact of a change in numbers on our qualitative impressions of a bird community. In the present chapter, therefore, we are concerned both with how the status of individual species may be linked with farm structure and with how the total numbers of birds on farmland may be determined.

Breeding habitat correlations

We have examined the correlations between the breeding densities and habitat features on each of 65 farmland CBC plots surveyed in 1965, when the population levels of many bird species were still low after the mortality of the 1962–63 winter – habitat preferences are best detected in such conditions (van Horne, 1983; O'Connor, 1986). We summed the densities of the various species breeding on each farm to estimate the total pairs breeding there. These overall densities are correlated with the local densities of hedges, small woods, and ponds (Fig. 6.2). All three correlations shown are mutually independent, even though farms rich in hedgerow tend also to be rich in ponds.

Fig. 6.2. Total density of breeding birds on farmland in relation to (top) hedgerow density, (middle) pond density, and (bottom) woodland density, on each farm.

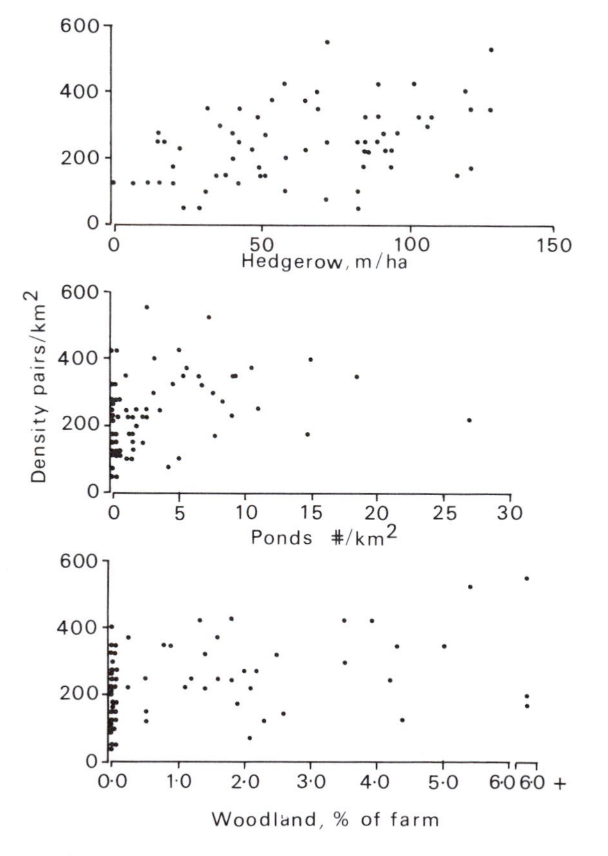

The number of species and the number of pairs on each farm are affected to different extents by different habitat features. In other words, increases in some habitat features result in additional species appearing but increases in other habitat features either introduce more individuals of species already present or introduce new species at the expense of those already present. For example, species totals decline with the intensification of tillage to a lesser extent than do breeding densities, probably because the loss of abundant hedgerow species such as Blackbird or Whitethroat is offset by the gain of less numerous field species such as the partridges or the Corn Bunting. Again, the amount of hedgerow on a farm does not influence the number of species found there but does sharply influence overall breeding density. In contrast, the local density of farm ponds is equally well correlated with density and with species count.

Table 6.2 extends this analysis to individual species and lists the preferred (i.e. most strongly correlated) habitat feature for the 57 commonest farmland species. The availability of hedgerows containing trees was the most significant landscape feature for 15 species (26 per cent); woodland density was next, affecting eight species (14 per cent); the proportion of fields given over to arable farming (including tillage) was next (six species); the presence of ponds and of scrub each affected five species; and the density of hedges on the farm affected four species. The remaining 14 species considered were variously linked with individual landscape features. The table of course represents only the crudest of partitioning of habitat preferences, given the intercorrelation of habitat features on farmland, but its general pattern is likely to be correct. This is suggested by the second part of the table which tallies the number of species showing any level (and not just maximum) of statistical correlation between the abundance of each species and each habitat feature. Again, the extent of hedges containing trees most frequently appears, followed by pond and woodland densities.

Several habitat variables – notably ponds, hedgerow density, openness, and field size – were correlated with markedly more species in the second tabulation of Table 6.2 than in the first. This may suggest that, whilst few species were wholly dependent on the presence of such habitat for breeding, many more species benefited where they were present. No less than 25 species were affected by the openness of habitat but only for Corn Bunting was this the maximum correlate. Similarly, although field size and hedge density (which, not surprisingly, are negatively correlated) each influence the abundance of 40 per cent or more of the species, their effects are swamped by other habitat features for all but four (7 per cent) of the species examined.

Wyllie (1976) undertook a similar analysis within the mainly cereal farmland of the parish of Hilton in Huntingdonshire. Fig. 6.3 summarises the spatial coincidence of territories with various habitat features there. The proportion of the community associated with hedgerow in his survey is larger than in our analysis but this is probably because woodland was essentially absent from his study area. The use of arable fields emerges more strongly in Wyllie's (territory-based) survey than in our (species-based) analysis, probably because his largely cereal area had relatively more Skylarks present (33 pairs per km^2 or 14.4 per cent of the community) than we had (Chapter 2). The influence of ponds and streams was rather

Table 6.2. *Numbers of species numerically correlated with various habitat features on farmland*

	Maximum correlations[a]		Species affected[b]	
Habitat variable	Number	%	Number	%
Hedgerow containing trees, m/ha	15	26.3	43[c]	75.4
Hedgerow without trees, m/ha	0	0.0	9	15.8
Hedgerow density, m/ha	4[d]	7.0	30[d]	52.6
Lines of trees, m/ha	3	5.3	20	35.1
Woodland, ha/ha	8	14.0	32	56.1
Scrub, ha/ha	5	8.8	20	35.1
Field size, fields/ha	0	0.0	24[d]	42.1
Arable fraction, ha/(ha of field)	6	10.5	18	31.6
Openness index %[f]	1	1.8	25[e]	43.8
Farmsteads, ha^{-1}	3	5.3	15	26.3
Ponds, ha^{-1}	5	8.8	33	57.9
Linear water, m/ha	3	5.3	16	28.1
Altitude, m	2[d]	3.5	11[g]	19.3
Slope, m/m	2	3.5	12	21.0

[a] Number of species more strongly correlated (Spearman rank correlation) with this habitat variable than with any other one. Correlations positive except where stated otherwise. A total of 57 species was analysed.
[b] Number of species showing statistically significant correlation (Spearman rank correlation) with the habitat variable.
[c] Two of these species were negatively correlated with the habitat.
[d] One species here was negatively correlated.
[e] Sixteen species gave negative correlations here.
[f] Percentage of overlaid grid points that did not fall on a three-dimensional habitat feature (e.g. hedgerow, tree, woodland, farmstead, etc.).
[g] Three species gave negative correlations here.

lower in his survey than in ours, again probably because ours encompassed a greater diversity of habitat.

Some 14 per cent of the species commonly or regularly found in the farmland CBC censuses are aquatic in habitat and are thus largely confined to available water (Table 6.2 above). Common Birds Census data tend to underestimate the importance of these birds on agricultural land, since some species, such as the Reed Warbler, are colonial nesters not readily censused by the territorial mapping of the CBC, and others, such as Mallard, have large ranges difficult to delimit. The presence of major aquatic features such as streams tends to overshadow the general value to birds of farm ponds. These emerged in our analysis as an important correlate of bird numbers (Fig. 6.2), with the abundance of five species maximally correlated with the habitat and with 33 species significantly influenced by pond numbers (Table 6.2). On many lowland farms ponds were established to provide a reliable source of water for farm stock. But water is also essential to many passerine birds, particularly in summer, and birds will travel quite long distances to visit a source of water. Ponds also provide sources of nest material for hirundines, summer feeding on their

Fig. 6.3. The proportions of birds correlated with hedges and other features of the environment in the parish of Hilton, Cambridgeshire, according to the study of Wyllie (1976). Territories were assigned to habitats on the basis of net location where known, otherwise of territory centre-point location.

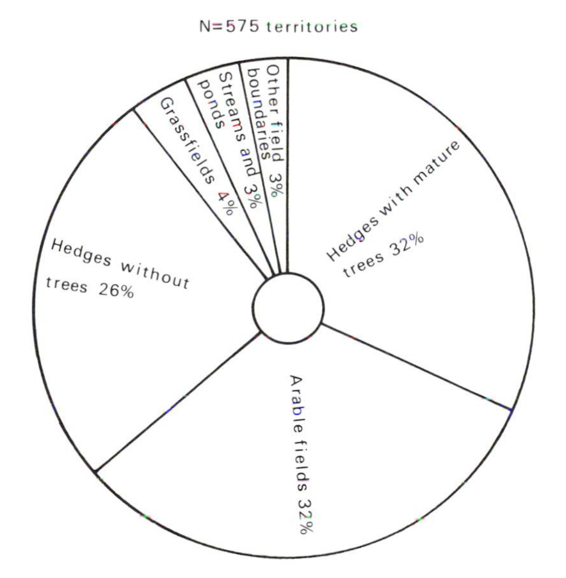

margins for species such as Song Thrush, and reliable source of food for Swallows (Møller, 1983). Ponds thus provide a general improvement to the farm environment rather than specific nesting habitat, and their value for farmland birds is probably under-appreciated.

Ponds provide good nesting habitat for certain species. The abundance of Moorhens was particularly associated with the spatial density of ponds on these farms and farms without at least one pond present were unlikely to have any Moorhens present (Fig. 6.4). Although Moorhens also nest along waterways, the extent of such habitats was not nearly as good a predictor of numbers as were ponds. Both ponds and rivers vary in the quality of territory they offer to Moorhens and in Denmark Jorgensen (1975) found that pond size influenced selection, though small ponds could be used by birds obtaining some of their food from the surrounding fields.

Fig. 6.4. The density of Moorhens on farmland in relation to the local density of ponds: (top) farms without ponds, (middle) farms with 1–5 ponds/km², and (bottom) farms with six or more ponds/km².

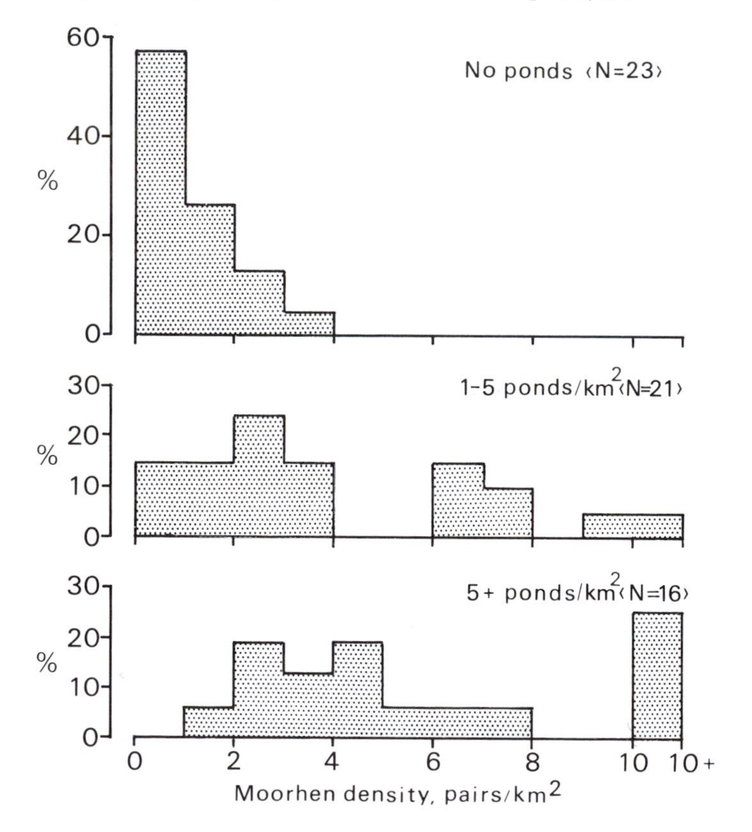

Data from a CBC plot in Hertfordshire show that making even a small pond allows Moorhens to colonise new areas. The selection of ponds by Moorhens is also influenced by the extent of aquatic vegetation present (Jorgensen, 1975). Similarly, where bankside and emergent vegetation are present along waterways breeding is both earlier and more successful (Taylor, 1984).

Mallard are more numerous on farmland the greater the length of river present. Ponds are less frequently used for breeding, perhaps because pond sizes are typically too small for Mallard (cf. Jorgensen, 1975). Bankside vegetation is probably critical for this species, the female nesting either in thick cover at ground level, or less frequently, in holes or hollows in mature trees on the bank. A number of passerines are directly associated with riverside habitat, the principal ones on lowland farms being Reed and Sedge Warblers, Pied and Yellow Wagtails, and Reed Bunting. Of these the Reed Warbler is strongly dependent on reedbeds and is readily displaced by dredging and bank-clearance operations (Williamson, 1971). Sedge Warblers are more catholic in their use of habitat but are still susceptible to river management. On one CBC plot in Kent, for example, maximum numbers of these two species occurred in years when dyke vegetation was left uncut. Williamson (1971) presents a case history for Sedge Warblers displaced from a stretch of the River Stour in Dorset: these redistributed themselves into field drains and wet ditches where the population persisted for several years, eventually being lost to ditch-clearance operations. In arable areas Sedge Warblers can, however, breed in such crops as rape or wheat if riverside vegetation is destroyed but some evidence suggests that their nests are less successful over dry land. Arnold (1983) concluded that ditches had rather little effect on breeding densities except where they bordered a hedge. Ditches may be more important as feeding habitat, for in Ireland Moles (1975) found that 18 species – principally Snipe, Moorhen, Blackbird, Song Thrush, Starling, and Reed Bunting – made significant use of them.

The migrant Yellow Wagtail is primarily associated with wet meadows along the banks of lowland rivers but also breeds in cereals and other crops on better drained land. These habits are reflected in our analyses in correlations of density with openness and with linear water. Its mainly resident congener, the Pied Wagtail, is one of the commonest riparian species and has apparently spread into agricultural land further away from water as a result of population increase (Williamson & Batten, 1977). Its density is correlated loosely with several different habitat features, which explains why it reaches greatest numbers on mixed farmland. Its winter distribution (Lack, 1986) is even more concentrated into areas of mixed farming, where strong territoriality along river margins may be present, with individual birds defending a stretch of river edge from which to harvest the invertebrates washed ashore (Davies, 1976). Birds unable to acquire territories feed in flocks, foraging over a variety of clumped food sources.

A number of riparian species make little use of the agricultural habitats surrounding them but can be affected by farming or farming-related operations. Thus Kingfishers need vertical banks for their nest tunnel and also need fishing perches over the water. Both features are destroyed where banks are graded and cleared of bankside trees in the course of land drainage (Williamson, 1971).

Much of what has been written about 'prairie' farms finds support in our data. Where such landscape elements as hedgerow, patches of scrub and trees are scarce or absent, birds are few, and where these elements are abundant, so too are birds (Fig. 6.5(*a*)). The number of species breeding

Fig. 6.5. Variation in (*a*) total breeding density, and (*b*) number of CBC species, in relation to habitat openness (percentage of sample points without three-dimensional habitat features such as hedges, trees, farmsteads, etc.). Vertical bars indicate ± standard error. Open dots indicate pooled data for values in the range 70–95 per cent.

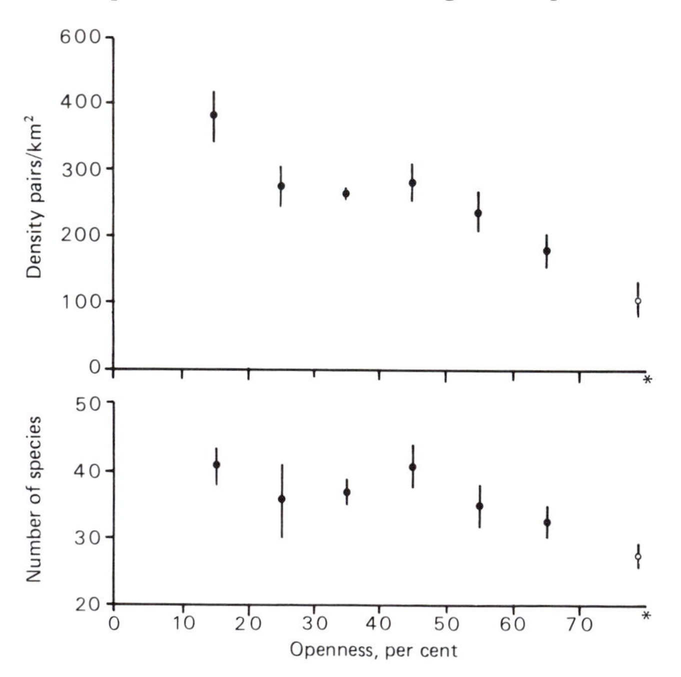

on a farm declines only slightly as the surrounding landscape opens out (Fig. 6.5(*b*)), probably because the loss of very numerous hedgerow species is partially offset by the gain of less abundant field species. Species present in open landscapes are thus represented by fewer individuals on average than are the species of mosaic landscapes. The Skylark is in fact the only common open country species whose farmland densities exceed 10 pairs per km^2. These generally low densities are due in part to lower tolerance of human disturbance (van der Zande, Kewis & Weijden 1980). Open country species in the Netherlands then fly in response to human disturbance at longer distances than do the more typically woodland species, possibly because they lack cover in which to take refuge, and they adapt to regular disturbance more slowly. They therefore avoid nesting in the immediate vicinity of roads and farmsteads and this restricts their density over the farm as a whole.

Farm roads may themselves constitute a habitat element permitting nesting or foraging by birds. Laursen (1981) found that although roadside verges accounted for less than 4 per cent of an average Skylark territory in his Danish study area, they accounted for nearly twice this proportion of foraging sightings. Similarly, four times as many Skylark nests were on verges as were expected on a *pro rata* basis. Laursen suggested that in early spring, when Skylarks choose their territory and nesting sites, roadsides were the main sources of green vegetation and cover because barley fields, which accounted for two-thirds of the crops in the study area, are not sown there until April and May. In addition, those birds which nest in barley or in beet fields may be disturbed by work in the fields until late May. In fact, all the Skylark nests found on verges were adjacent to fields with spring-sown crops. In East Anglia Skylarks tend to use tall vegetation at the edge of cropped areas and are scarcer on arable land lacking the ditches or hedges that promote such growth (Arnold, 1983).

Farmsteads similarly have opposing effects on birds. Disturbance due to work around farm buildings depresses breeding densities on the surrounding land. In van der Zande *et al.*'s (1980) study of Lapwing and Black-tailed Godwit, for example, breeding densities regained their undisturbed level at 220 m from an isolated dwelling house but the disturbance distance extended to 470 m in the case of an isolated farm and its associated operational activities. Ground-nesting birds may also avoid the cats and dogs present around most farmsteads. On the other hand, the farm buildings themselves may provide nesting niches not available elsewhere on the farm. Swallows are essentially dependent on farmsteads in this way and population densities on individual farms increase with the area of farm buildings present (Møller, 1983). Such species as Barn Owl and Kestrel, as well as the commensal House Sparrow, may be only slightly less dependent on such buildings. On intensively arable or 'prairie' farms, moreover, the gardens and hedges around the farm centre are effectively islands of the sole available habitat for many species. Shrubb (1970) estimated that farmsteads and gardens formed only 2 per cent of the total area of his (arable) Sussex farm, yet held 20 per cent of the total hedgerow on the farm and housed 46 per cent of all hedgerow and garden birds. Such hedgerows are in addition less likely to be removed than are other internal hedgerows on the farm and are more likely to hold mature trees. The densities of several species normally associated with mature trees – among them Stock Dove, Tawny Owl, Great Spotted Woodpecker, Jackdaw and Redstart – were correlated with the density of farmsteads on each plot (O'Connor, Morgan & Marchant, in prep.). Farmsteads also serve as an important source of winter food for many resident passerines (Chapter 4).

Lines of trees without an underlying hedge are a feature of some farms in Britain, particularly in East Anglia. Only two species – Great Spotted Woodpecker and Spotted Flycatcher – were numerically related to this habitat in our sample. Such lines of trees were predominantly relict and associated with arable farming and many trees were probably dead or dying. The association of Great Spotted Woodpeckers with such conditions is to be expected and was only slightly stronger than with woodland generally. In contrast, the woodpecker was only moderately associated with the hedgerows containing trees, presumably because these were in better condition, containing less dead or dying wood. For Spotted Flycatcher both lines of trees and hedgerow trees offer foraging perches but the former provide a more open sallying area.

The relative importance of woodland and hedgerow in maintaining the bird communities on farmland has only recently begun to receive systematic study. The idea that tree plantings are an effective replacement for the hedgerows grubbed out during farm modernisation has been put most forcefully by Murton & Westwood (1974). It is based on the notion that hedgerows are suboptimal habitat for farmland birds of woodland origin which overflowed into the newly created hedges of the Enclosures only under the impetus of population pressure. Osborne (1982a) has argued strongly that this view should not guide conservation strategies, noting that, whilst hedgerows may indeed hold smaller densities than would the same area of woodland, the sheer volume of agricultural land in Britain results in many more birds in total living in hedgerow than in woodland. O'Connor (1984a) provides some evidence in support of this view but there are several practical difficulties in assessing the relative importance of hedgerow and woodland, not least the possibility that birds may respond to the matrix of woodland and hedgerow rather than to the absolute abundance of either. Indeed, hedgerow nests immediately adjacent to woodland have been shown to be less successful than nests further away from the wood, apparently because woodland predators will not move far from the wood when searching hedgerow (Osborne, 1982a). Similarly, hedgerow may be prime habitat on farms where woodland is scarce and yet be quite unimportant where alternative habitat is plentiful (Fuller, 1984). O'Connor (1985) reviews the factors likely to influence the relative use by birds of optimal and suboptimal habitat.

Such factors may also be relevant in considering the way in which birds other than prairie species use fields. Although Fuller (1984) found that few birds used fields on Manydown Farm in Hampshire, at Oakhurst many species, but particularly passerines, range into fields to exploit farmwork or a particular stage of crop development. This difference between the two farms is probably related to the presence of some 62 ha of woodlands on the Hampshire farm, covering some 11 per cent of the land; at Oakhurst such habitat comprises only 1.5 per cent of the 173 ha. The finding that behaviour towards one resource may be modified by the other resources

Fig. 6.6. The relationship of species richness to woodland abundance on farms (*a*) with below average density of woodland, and (*b*) with above average density of woodland. Data from 65 CBC plots surveyed in 1965.

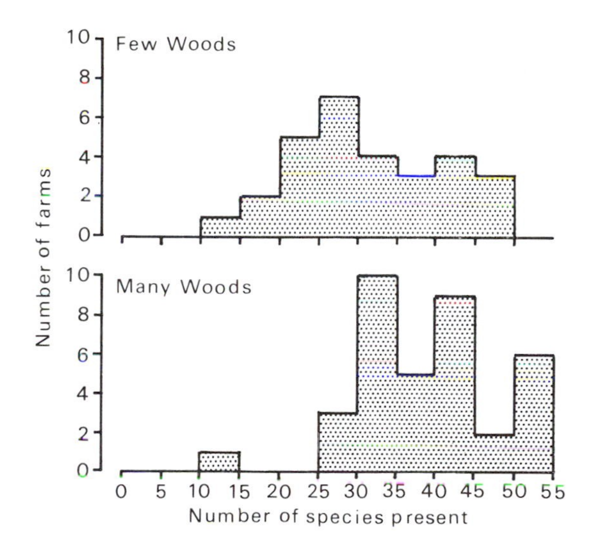

available on farmland is an important caveat and may explain some of the anomalies in previous reports on farmland habitat use.

Our own data show that the absolute density of woodland on a farm – irrespective of any consideration of its spatial patterning – is correlated with the number of species breeding there (Fig. 6.6). Each additional species is represented only by rather few individuals, whilst additional hedgerow not only increases species totals but brings in more individuals of each new species. This might seem to indicate that extra hedgerow is better for birds than is extra woodland but one must recognise here that our results were established against a background of existing woodland. On the farms we examined, more hedgerow may be of greater value than additional woodland but on other, less wooded, farms this might not be so. We show in the next chapter, for example, that bird abundance does not always increase linearly with habitat abundance. Similarly, Fuller (1982) has shown that the various species found in woodland in Britain have characteristic area thresholds in respect of the woods in which they breed. The Tawny Owl, for example, is absent from fewer than 10 per cent of all woods above 80 ha but is absent from nearly half of all woods below 10 ha in size.

Additional woodland will increase the breeding density of at least some of the species already present on a farm, as shown for Jackdaw in Fig. 6.7. However, as expected from the previous paragraph, such relations are by no means general and our analyses show that the densities of only eight of some 57 species examined are in fact most closely correlated with farm woodland (against 19 best correlated with hedgerow). The abundances of a further 24 species are also correlated to some extent with woodland density, though they have still stronger correlations with other habitat features. This could in principle be an artefact of cross-correlation amongst

Fig. 6.7. Density of Jackdaws in relation to woodland on CBC farmland plots.

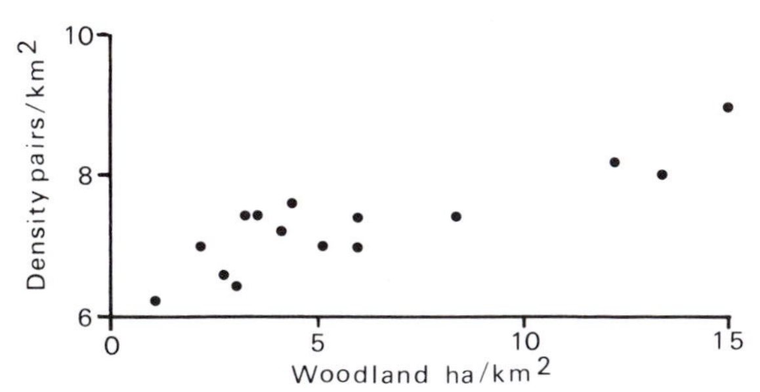

our variables (e.g. if the farms with most hedgerow were also those with most woodland) but in practice the co-occurrence of wood and hedge or similar pairings is surprisingly small in our samples. It follows that the birds concerned are able to use one habitat as a substitute for the other. But it also means that more birds can breed on farms where the overflow habitat is available than where it is not available. What we really need to know, therefore, is not whether hedgerow or woodland is important for birds but rather just how much woodland is needed by way of replacement for lost hedges. This remains a problem for the future.

Much of the above (and, indeed, much of the literature on this subject) assumes that a bird's response to any one habitat feature is not affected by other habitat features present. However, Williamson (1971) emphasises the importance of maintaining a hedgerow system in continuous contact with copses and small woods from which birds will disperse. He concluded that Wren and Chaffinch each used farmland hedges as a result of overflow from woodland centres of population. A similar effect is described by Bull *et al.* (1976) of Robins and by Osborne (1982a) of Chiffchaffs (see Fig. 6.9 below). Rather little research has been done on this topic, a serious omission given the views of some conservationists that the planting of odd field corners can provide sufficient habitat to accommodate the birds found in hedges. What is really needed to test Williamson's argument is a comparison of the distribution of birds over farm copses on farms respectively with and without linking hedgerow.

Winter habitats

In contrast to breeding habitat requirements, habitat use by birds in winter is relatively under-studied. Fig. 6.8 shows that winter and summer habitat use by birds on farmland in Co. Down (Northern Ireland) differed principally in respect of greater use in winter of marginal open ground and permanent grass and of poor quality hedgerow. Grassland feeding by Starlings and Chaffinches in winter accounted for a large part of the winter use of the former whilst hedgerow foraging by Blue Tit and Dunnock accounted for the latter. Presumably such low thin hedges broken by many gaps simply do not offer enough cover for nesting during the summer. The figure also shows that metalled paths are used in summer but not in winter, mostly by Dunnocks and House Sparrows. Greater invertebrate activity – conspicuous against the background of the path – or a need to collect grit for nestlings are possible explanations. Moles' (1975) data on the greater importance of grassland in winter are supported by Arnold's (1983) study in East Anglia, where more species were present on grassland in winter than on arable land with the same habitat characteristics. Grassland with hedges and trees was particularly favoured by immigrant thrushes and buntings, especially on those plots where woodland was scarce or absent. Resident thrushes, Wrens, and Dunnocks were very scarce on arable land where hedges were low but the presence of a tall hedge or (in the cases of the last two species) linear woodland resulted in increased numbers. Ditches, particularly if of large volume, were important feeding areas in winter, especially for Blackbird, Song Thrush, Robin, Wren, Dunnocks, and buntings, again as in Ireland (Moles, 1975).

The nature of the surrounding countryside has a marked effect on winter bird numbers, especially where hedgerow is scarce (Arnold, 1983). The extent of gardens available nearby is significant, presumably either because of their value as refuge areas in severe weather (van Balen, 1980; O'Connor, 1980; Spencer, 1982) or because they serve as centres from

Fig. 6.8. Relationship between habitat utilisation (all species pooled) in summer and in winter on farmland in County Down, Northern Ireland. Each dot represents one habitat element. Habitats equally favoured in winter and in summer lie along a line 45 degrees to both axes. The lettered dots refer to habitats discussed in the text: A, marginal open ground; B, marginal permanent grass; C, low thin hedges with gaps; D, metalled path. Data from Moles (1975). Utilisation was measured as registrations per 300 m² of the habitat over five censuses.

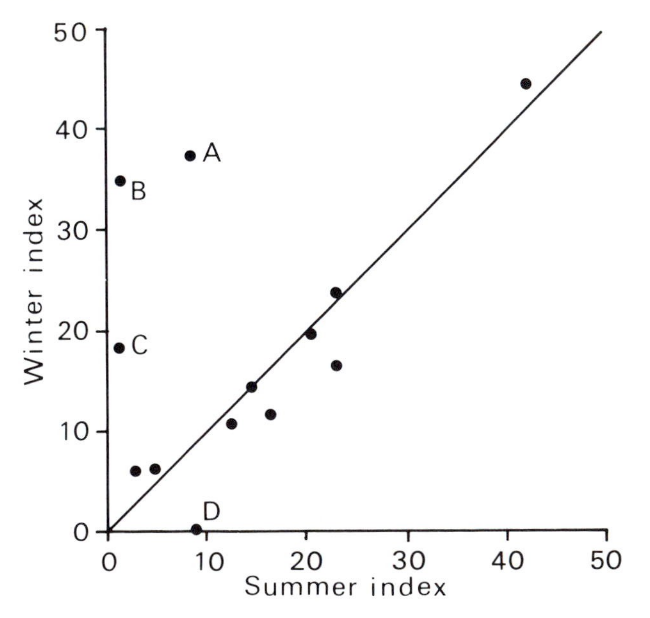

which to forage out into farmland (Arnold, 1983). In winter a number of species such as titmice prefer to feed in trees and in East Anglia these species fed in hedgerows in inverse proportion to the local extent of woodland and lines of trees. Lines of trees were largely absent from Moles' Irish study area. The marked dependence of winter bird numbers on the characteristics of the general countryside, combined with the individual species preferences for trees, ditches, grass, and so on, explains much of the correlation between winter bird diversity and agricultural diversity established in Chapter 2 (Fig. 2.10).

Are farmland habitats full?

Implicit in the use of correlations between breeding densities and habitat features is the idea that more habitat means more birds. In reality the available habitat is usually (in Britain) divided amongst birds by their territorial behaviour and this does not always result in even division of resources. Territorial behaviour usually reserves a supply of food or a foraging area from which to feed young (e.g. Stenger, 1958; Schoener, 1968; Gill & Wolf, 1975). When population density is low, most birds defend large territories but at higher densities the amount of defensive behaviour necessary increases and the length of defended boundary is reduced but defended with greater vigour. In this way resistance to further reduction in territory size increases and a lower limit to size is set. Even with such a limit, clutches are often smaller in densely populated habitats, since each pair has a smaller area over which to gather food for their young (O'Connor & Fuller, 1985). In other species predation risk can be reduced if the nests are spaced further apart than at random, as happens with territorial spacing. The net effect of these various factors is that density may alter with the extent of a particular habitat in a non-linear manner but eventually the habitat is filled completely. Further would-be settlers are then excluded from their desired habitat but may breed elsewhere in poorer habitat, as in Great Tits (Krebs, 1971) and Woodpigeons (Murton & Westwood, 1974).

Williamson (1969) examined habitat use by the Wren population of English farmland as it recovered from the cold winter of 1962–63. He used data from nine farms, selected partly for the variety of habitat they afforded. In 1964 and 1965 territories were established primarily in wooded areas or in vegetation alongside rivers and streams, but as the population continued to increase and the woodland and streamside areas filled up, proportionately more of the territories were based on gardens and orchards. Late in the period, in 1966 and in 1967, there was a marked increase in the use of field and lane-side hedges. Territory size in woodland areas was fairly constant, irrespective of population size, typically occupying one to one-and-a-half acres. In hedgerows, on the other hand, linear territories usually extended for one-fifth or one-sixth of a mile, often with a concentration of activity at hedgerow intersections. The Wrens also concentrated into those hedges most rich in shrub diversity (Osborne, 1982a). Habitat selection by other farmland species also varies with local population density in this way. Fig. 6.9 shows this for the population of Chiffchaffs breeding on a Dorset farm: at low densities all the territories were centred on small woods and copses but, as densities increased,

Fig. 6.9. The distribution of Chiffchaff registrations from mapping censuses in relation to the positions of copses on farmland, showing how distant hedges were used only in years of high population level. From Osborne (1982a).

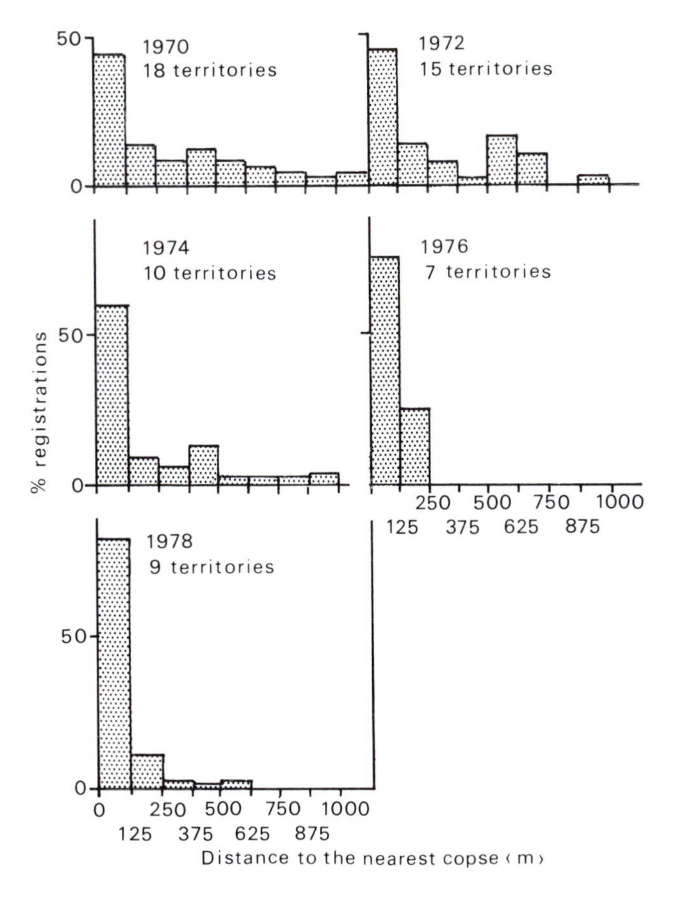

relatively more individuals were forced further away from the preferred habitat.

Williamson noted the relative instability of Wren territories in the less favoured habitats. He found that a higher proportion of territories in such habitats failed to be reoccupied the following year than was the case with woodland sites. On his nine farms he found only one 1964 territory (2 per cent) was left vacant in a subsequent year, whereas of the new territories first established in 1965 some 12 per cent were subsequently left unoccupied and of the 1966 territories 22 per cent were unoccupied in 1967. Although some mortality would be expected, Williamson considered that all good habitats would have been colonised by prospecting males each spring, and attributes the changes in dispersion rate to a shifting of bird territories as birds move from poor to better habitats. Both Wren and Robin populations have been shown to breed in hedgerows when the available woodland is fully stocked and to abandon them as quickly whenever a fall in breeding numbers creates vacancies in the woodland (Benson & Williamson, 1972). Such dynamical shifting in habitat spectrum is common in birds (Fretwell & Lucas, 1969; Newton & Marquiss, 1976; O'Connor & Fuller, 1985; O'Connor, 1986).

Other evidence showing that access to suitable habitat may be limiting the number of birds able to breed comes from an elegant experiment with Hooded Crows in Scotland (Charles, 1972). In the open farmland around Newburgh, Aberdeenshire, most Crow territories are centred on small clumps of trees in which territory owners can nest but to which other birds are denied access. When a tree of suitable size was experimentally imported onto the boundary between two existing territories, a pair of birds previously restricted to the non-breeding and non-territorial flock were able to centre themselves on this tree and create a new territory from the fringes of the original territories. In addition to such studies, the older literature contains many anecdotal accounts of long series of one bird of a nesting pair being shot, yet being replaced almost immediately by a new individual, again indicating that more birds may be seeking to establish themselves than can be accommodated in the habitat available. Holyoak (1974) estimated that about 34 per cent of the spring population of Carrion Crows in southern England were non-breeders, in Choughs the figure was at least 30 per cent and in Magpie as many as half the spring population were non-breeders, though holding territories. Amongst songbirds, the Blackbird has a particularly large non-breeding population, with about 40–45 per cent of males lacking territories (Batten, 1977; Edwards, 1977).

Territorial behaviour of this type has several consequences for the dynamics of bird populations on farmland (Williamson & Batten, 1977). Where farmland is a less preferred habitat, the population density of the species concerned varies more on farmland than in woodland. Whenever vacancies in woodland occur, they are filled from the reservoir of farmland birds, thus keeping the density in woodland relatively constant at the expense of that on farmland. The only time woodland densities of these species decrease significantly is after an extremely severe winter, when survivors are too few to fill the available woodland vacancies. For this reason the CBC indices for several common farmland species show a general pattern of stable woodland populations with decreases apparent only after severe winters, whilst farmland populations are unstable and marked with even larger deviations due to severe winters. Such patterns imply that woodland populations are nearly always at full carrying capacity whilst farmland numbers are nearly always below capacity, but we question this conclusion for two reasons.

Our first objection lies with the major consequence of agricultural crop management in altering the form of the niches available on farmland at any one time. The effects of this on the extent of apparent under-use of farmland by birds have not hitherto been considered. Consider the simple case of a ground-nesting species requiring spring tillage as a source of food. For such a species any variation in the proportion of spring-sown and autumn-sown crops will alter the amount of such tillage and therefore the number of pairs that the farm will support in any one year. Conversely, some other species may respond in the opposite way (because of its dependence on late summer grain, for example) and thus have higher numbers in years of much autumn-sown crops. In such conditions the summation of the peak counts across years will not reflect the carrying capacity of the farm. For hedgerow species such changes may be less significant when estimating levels of farm use, except where particular crops are used as foraging areas.

Second, much of the detailed evidence of under-use of farmland is equivocal. In a series of case studies of the carrying capacities of farms of different types Williamson and his collaborators estimated that farmland habitat was rarely filled to saturation, the extent of the underoccupation being rather consistently around 20–25 per cent. These estimates were made by summing the maxima for each species observed over a long run of censuses to get a hypothetical 'best season' for the farm. On this basis Bull *et al.* (1976) estimated for their Norfolk farm that the normal exploitation rate was around 75–80 per cent of the potential maximum. In Suffolk the figure varied from 71 per cent in 1964, when the population had not fully recovered from the 1962–63 winter, to 85 per cent in 1968, with a long-term average of 78 per cent (Benson & Williamson, 1972). If underexploitation of farmland habitat is genuinely at this level, it suggests that there are always some unoccupied niches for some or all of the bird species of the farmland community and farms should be able to withstand a degree of habitat modification or loss without visible adverse effects of the breeding bird community. In practice this does not appear to hold. On the Norfolk farm already mentioned, for example, structural changes resulted in losses of Wren (38 per cent), Blackbird (25 per cent), Dunnock (20 per cent) and Song Thrush (19 per cent).

As farmland is a recent habitat in evolutionary terms, it is unlikely that its woodland colonists have yet fully evolved the isolating mechanisms needed for full coexistence (Lack, 1971). That is, farmland may be seen by birds (in the absence of effective isolating mechanisms) as consisting of a number of generalist niches that may be filled *either* by species A *or* by species B. In such conditions a farm with, say, 10 niches might be occupied in one year by seven pairs of species A and by three pairs of species B but in a different year by three pairs of species A and seven pairs of species B. The method of Williamson and his colleagues used on such data would indicate a carrying capacity of 14 pairs instead of the true level of 10 pairs. Such 'generalist' niches do appear to be present in farmland in winter (Green, 1978). Skylarks, Woodpigeons, and Grey and Red-legged Partridges all have very similar seasonality in diet, turning from an early use of cereal grain to grazing in late winter. Overlap in winter diet is also common in finches (Newton, 1967). Some evidence of interspecific competition on farmland during the breeding season is also available, for migrant species nesting on farmland occupy narrower nesting niches when resident populations are abundant and occupy wider ones when the resident populations have fallen after a severe winter (R. J. O'Connor, unpublished). Similarly, the three congeneric buntings commonly found on CBC plots have quite different habitat spectra (O'Connor & Fuller, 1985). Corn Buntings are most numerous in the most prairie-like farms; Yellowhammers show exactly the opposite correlates; and Reed Bunting numbers are correlated with the amount of aquatic habitat on the farm. Such habitat segregation often reflects competition (Lack, 1971). Fig. 6.10 illustrates this for the various buntings present on a Yorkshire CBC plot surveyed over 23 years. All three species fluctuated in numbers but the peaks are out of phase, those for Yellowhammer and Reed Bunting especially moving in opposite ways. In other words, when the farm supported many Reed and Corn Buntings it held few Yellowhammers, and conversely. The maxima of three species were 15 Yellowhammer, 14 Reed Bunting and four Corn Bunting pairs respectively, for a grand total of 33 pairs, yet at no one time were there more than 16 territories present.

These two points remove much of the force behind the idea that farmland habitat is under-used by birds. The various correlations presented

Fig. 6.10. Variation in the numbers of three bunting species on a Yorkshire CBC plot 1961–1984, showing the complementary trends between Yellowhammer abundance and that of Reed and Corn Buntings. No censuses were conducted in 1967, 1976 and 1981.

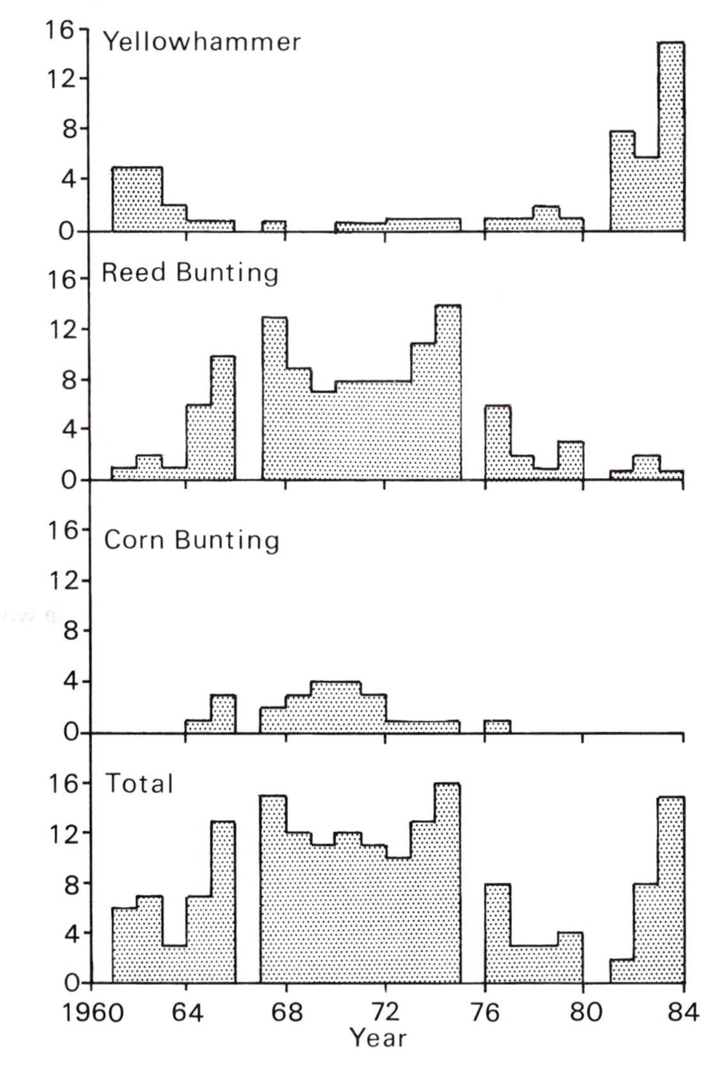

earlier argue for a direct link between farm structure and carrying capacity for birds, particularly as these correlations were measured in a year when population densities were low and the birds relatively free to distribute themselves into the best available habitats. However, densities do not always respond *pro rata* to changes in habitat density, and also habitat use may alter with population pressure; this means that density distribution does not always equate with habitat preference. These points have critical implications for habitat conservation, particularly in relation to the controversial issue of hedgerow removal to which we turn in the next chapter.

Summary

The majority of changes to farm structure nowadays involve habitat loss. The extent of hedgerow, woodland, and farm ponds on individual farms each influences the density and variety of birds breeding there. Individual species display a variety of habitat responses, including non-linear and compensatory ones, though practically all species are least abundant on the most prairie-like farms. Farm gardens are particularly rich in birds. The relative importance of woodland and hedgerow remains to be established, though in our sample additional hedgerow appeared more valuable than additional woodland. Winter habitat use has been poorly studied but grassland, ditches and gardens appear to be particularly valuable. Considerable evidence exists to show that some habitats are saturated at high densities of species, and the idea that farmland is normally merely a partially filled, overflow habitat for woodland species is by no means proven.

7: The effects of hedges and hedgerow loss on farmland birds

Birds and hedges have been studied over many years (Alexander, 1932; Chapman, 1939; Campbell, 1953; Williamson, 1967, 1971; Parslow, 1969; Hooper, 1970; Wyllie, 1976) but little attempt has been made to establish statistical relations between the two. In the previous chapter we showed on statistical grounds that hedgerow is a significant habitat factor influencing the size and make-up of farmland bird communities. The relative importance of hedges for birds has been controversial, however, particularly since the appearance of the paper by Murton & Westwood (1974). Most naturalists have opposed hedge destruction and have equated hedge losses with bird population losses, it being obvious to all that birds live in hedges. The immediacy with which hedgerow removal is apparent in a landscape no doubt has fuelled the arguments here. Murton & Westwood (1974) challenged these assumptions and, on the basis of a major case study, dismissed the issue of hedge removal as a red herring where farmland birds are concerned. Their arguments have regrettably in turn been elevated to the status of gospel, leading to serious advocacy of such notions as the equivalence of field corner plantings to the total hedgerow network. Recent research has shown that the truth is, as usual, somewhere between these two extremes (Osborne, 1982a; O'Connor, 1984a). We therefore review this issue in greater detail than we have afforded other habitat elements.

The number of species breeding on a farm does not vary linearly with the amount of hedgerow present but goes through a shallow peak at hedge densities of around 7–11 km per km^2. The number of pairs breeding, however, steadily increases with increase in hedgerow abundance, as shown in the previous chapter (Fig. 6.2). Thus the main effect of adding or removing hedges on a farm is to modify the local density of birds rather than the number of species present. This accounts for the anomaly reported by Murton & Westwood (1974), that the number of breeding species on their study farm at Carlton, Cambridgeshire, showed a net increase between 1960 and 1971 as two-thirds of the hedgerows there were removed. In detail six species were lost and nine gained, and 26 (of 66) species decreased in numbers against 17 increasing. What happened at Carlton during modernization, therefore, was that hedgerow density on a hedge-rich farm was moved closer to the optimal value for species total but at a cost in total breeding density. In contrast, the site studied by Evans (1972) began with only 5.3 km of hedge per km^2 and lost more than half of its breeding birds as hedges were removed down to a density of 0.28 km per km^2. This would be expected in the light of our finding of that most species were present at hedge densities of 7–11 km per km^2.

Influence of hedge structure and composition

Since the breeding bird community of woodland is heavily influenced by the structure and composition of the tree species present (Fuller, 1982), the detailed structure and composition of hedges are probably also important in determining what species are able to breed there. Were all hedges of standard size, shape, and species composition, there would be little opportunity for birds to discriminate between them. Hence we need to consider just how hedges vary on farmland. Helliwell (1975) examined the structure of roadside hedges in a Shropshire farmland sample, using coefficients of variation (standard deviation as a percentage of the mean) as a measure of variability in structure. He found that four factors were particularly variable, namely the number of trees per unit length, the number of shrub species in the hedgerow, the density of woods in the surrounding square mile, and the distance of the hedgerow to the nearest wood. The other measures he examined were rather less variable, though this would not necessarily be the case elsewhere.

The influence of hedgerow trees – the most variable component of hedges in Helliwell's (1975) survey – is particularly strong. Fig. 7.1 demonstrates how the distribution of breeding birds coincided with the distribution of trees and hedges in the parish of Hilton (Huntingdonshire), with high concentrations of birds along field boundaries and in the more wooded village area (Wyllie, 1976). Similar findings have been obtained elsewhere in Europe (Laursen, 1980). For our own data, the analysis of Table 6.2 in the previous chapter showed that for a majority of the typical 'hedgerow' species (including Blackbird, Wren, Robin and Whitethroat) the presence of hedgerow trees was significant. Indeed, the extraordinary feature of Table 6.2 is the total absence of any species most strongly dependent on hedges without trees.

Two case studies of the same Dorset farm, which was particularly rich in hedgerow trees but also included a significant proportion of low-trimmed hedge without trees, are given by Williamson (1971) and Osborne (1982a, 1984) and emphasise these points. Williamson found that large double-banked hedges with large mature trees (mostly elm but including oak, ash and willow) supported a bird territory every 32 yards (29.3 m) compared

Fig. 7.1. The distribution of birds in relation to the location of hedges and trees in the parish of Hilton, Huntingdonshire. From Wyllie (1976).

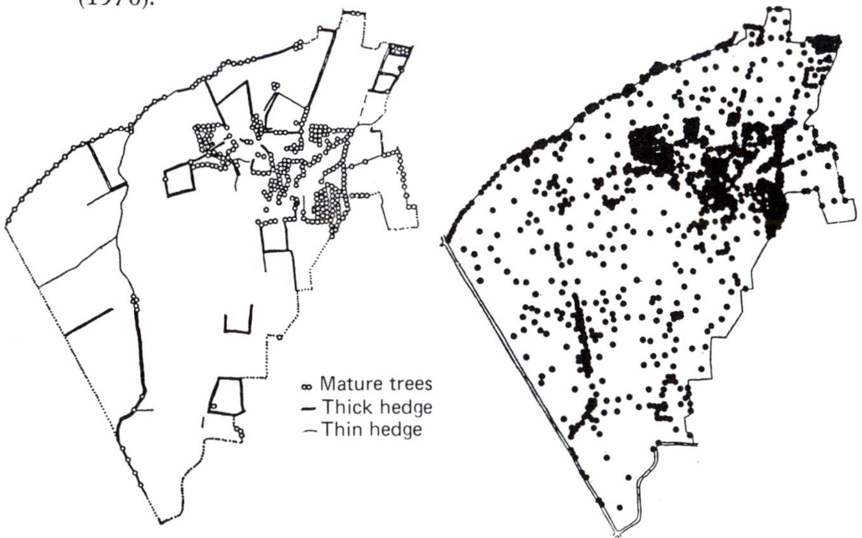

to one every 54 yards (49.4 m) in the low-trimmed hedges. The densities of the principal species along the two types are summarised in Table 7.1. The relative importance of the different types of hedges and their significance in relation to other habitats available on the farm are summarised in Table 7.2. The data show that 54 per cent of the main species and 43 per cent of all the territories registered were associated with internal hedgerows with trees, with a further 4.3 per cent of the territories centred on low-trimmed hedges. All species using the trimmed hedges (Whitethroat, Dunnock, Yellowhammer, Blackbird, Blue Tit, Wren, Sedge Warbler, and Linnet) were also found in the internal hedgerows with trees. The leading species in the latter are similar to those in the other well-hedged farms studied by Williamson (1967) in the Midlands and South of England, except that Chaffinch and warblers were unusually numerous in Dorset, the latter averaging 58 pairs of eight species. This was partly because a river attracted Reed and Sedge Warblers but also because the double-banked and ditched hedges of hawthorns and blackthorn interspersed with tall trees attracted Blackcaps, Garden Warblers and Lesser Whitethroats. On East Anglian farms warblers were likewise more numerous where such linear woodland was present (Arnold, 1983).

Osborne (1982a) studied the same Dorset farm as Williamson but after Dutch elm disease had infected the trees. He found then a net loss of species

Table 7.1. *Density of the principal species nesting in different types of linear habitat on a Dorset dairy farm*

	Length of hedge (m) per territory	
Species	Well-timbered internal hedges	Low-trimmed hedge and linear scrub
Wren	300	865
Dunnock	300	246
Robin	356	—
Blackbird	187	346
Song Thrush	855	—
Sedge Warbler	—	865
Whitethroat	—	246
Chiffchaff	430	—
Blue Tit	430	—
Great Tit	465	—
Chaffinch	262	—
Yellowhammer	—	246

From Williamson (1971).

from hedges following the tree removal, with the loss twice as large (24 per cent against 12 per cent) in those hedges losing half or more of their trees. Fewer Stock Doves and Magpies used the hedges following tree removal and the hedges that continued in use were those losing fewest trees. Long-tailed Tits also disappeared from the farm hedgerows, apparently because nest sites in the hedgerows became less suitable when greater penetration of light followed removal of diseased trees.

The historical origin of hedgerow trees in Britain is of some interest in the light of these results. Some were deliberately planted with the hedge itself, to create a coppice-with-standards approach to timber production. Elm, ash and oak were especially popular for this purpose during the main enclosures between 1760 and 1820 when the Industrial Revolution coincided with British involvement in wars in France, North America and India. Elm in particular was widely used in aquatic situations – for lock gates, for water wheels, and, together with oak, for shipbuilding. Other trees arose, however, simply by failure to cut them down when young, and allowing them to develop to maturity (Pollard *et al.* 1974). By either process the presence of trees tended to diversify the hedge itself as it aged, for ash regenerates from its tree stumps and elm produces suckers profusely.

Table 7.2. *Distribution of species totals and territories over different habitats on a Dorset dairy farm*[a]

Habitats[b]	Species N	Species %	Territories N	Territories %
Internal hedgerows with trees (6000 m)	25	(54.3)	206	(42.6)
Low-trimmed hedges (1000 m)	8	(17.4)	21	(4.3)
Gardens, orchards and farmstead[c]	23	(50.0)	69	(14.2)
Scrub (disused railway cutting) (700 m)	14	(30.4)	24	(5.0)
Relict natural woodland (elms) (1.25 ha)	25	(54.3)	54	(11.2)
Coppice woodland (ash, hazel, oak standards, hawthorn plus willows) (8.0 ha)	26	(26.5)	87	(18.0)
Riverbank (2340 m)	9	(19.6)	14	(2.9)
Totals	46	(100.0)	484	(100.0)

[a] In the analysis a few species, such as Mallard, Partridge, Pheasant and Stock Dove, are excluded on the basis of insufficient information and the Skylark is excluded because it is confined to fields.
[b] The extent of each habitat is given in parentheses.
[c] Extent not recorded.
After Williamson (1971).

Indeed, one of the consequences of the cutting-down of trees killed or made dangerous in the 1970s epidemic of Dutch elm disease has been a wealth of new suckers in many of the hedges affected.

Some idea of the size of the hedgerow tree population can be obtained from the Forestry Commission's 1980 survey which estimated 62.4 million hedgerow and parkland trees for England alone. Of the trees in use by nesting birds (as shown by the nest record cards), elm, oak and ash rank as the most frequent, though this may reflect the relative abundance of these species as much as any preference by birds. However, oak and ash are notable for their cavity formation (Burton & Osborne, 1980) and so for cavity breeders they are good alternatives to elms lost to Dutch elm disease. They are also amongst the tree species mostly widely in use by county councils replanting after Dutch elm disease (Osborne & Krebs, 1981). Pollard *et al.* (1974) quote figures for the hedgerow tree community prior to Dutch elm disease of 43 per cent ash, 25 per cent elm and 23 per cent oak, results based on a sample biased (in common with the CBC scheme) toward the south and east of England. The Forestry Commission census of 1951 gives a figure for elm of 21 per cent but by 1980 elms constituted only 3 per cent of volume for non-woodland trees. In Wales the corresponding figures were 13 per cent and 2 per cent respectively.

The number of shrub species was the second most variable feature of the hedges studied by Helliwell (1975). Floristic richness proved to be a

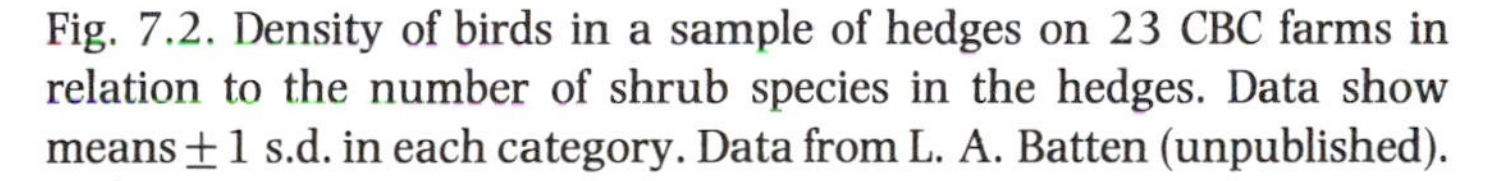
Fig. 7.2. Density of birds in a sample of hedges on 23 CBC farms in relation to the number of shrub species in the hedges. Data show means ± 1 s.d. in each category. Data from L. A. Batten (unpublished).

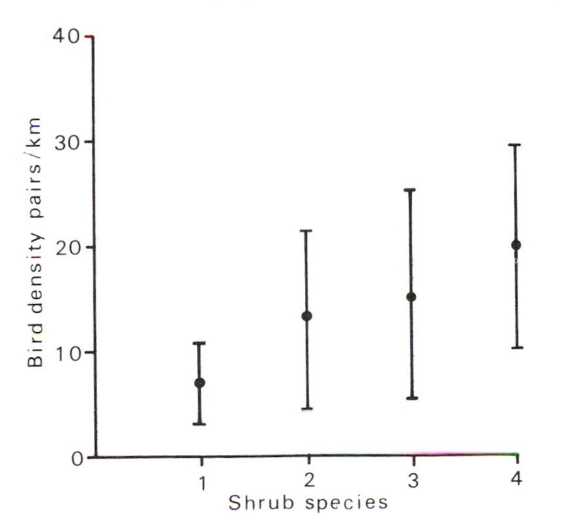

predictor of the Hertfordshire hedgerow bird communities studied by BTO staff (Fig. 7.2) and of the East Anglian farmland communities studied by Arnold (1983): more birds were present as the variety of shrub species present increased, though there is much scatter because other variables also have an influence. Farm hedges used by breeding Wrens in Dorset similarly contained a greater variety of shrub species than did the unused hedges (Osborne, 1984). The specific identity of the shrub species in each hedgerow also influences the bird community there. An unclipped but stockproof hawthorn hedge in the Midlands will average about 16.4 pairs

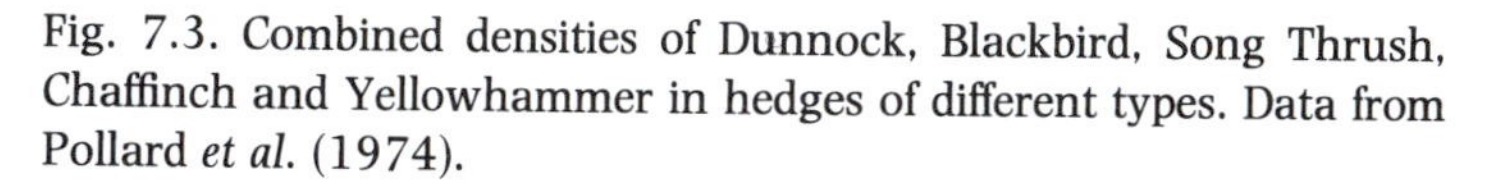

Fig. 7.3. Combined densities of Dunnock, Blackbird, Song Thrush, Chaffinch and Yellowhammer in hedges of different types. Data from Pollard *et al.* (1974).

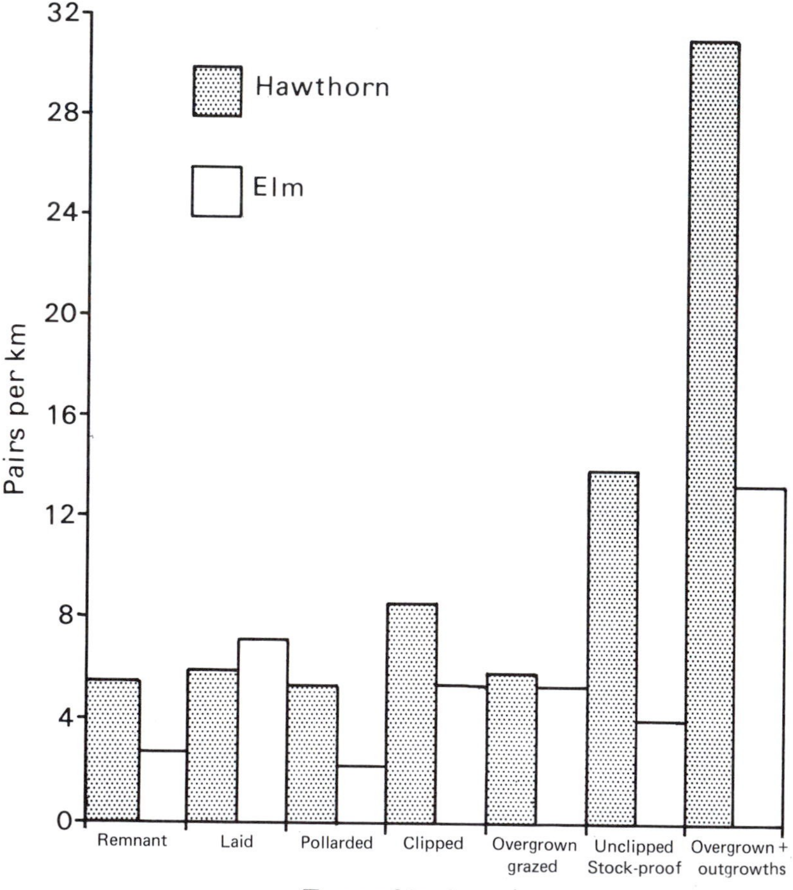

per km whilst a similar hedge of pure elm averages only 4.8 pairs. Even the best elm hedge types (neglected and unbrowsed) carry only about 15.3 pairs per km against the 37.2 pairs of hawthorn. There are several reasons why hawthorn might be preferred to elm, the principal ones being its denser cover, its earlier flowering (two or more weeks before elm) and its more abundant insect fauna (Pollard *et al.* 1974; da Prato, unpublished).

Structural changes brought about by hedgerow management are also important in determining the extent of hedgerow use (Fig. 7.3). Neglected hedges containing outgrowths of blackthorn and briar and with heavy field layers of bramble, nettles, and so on are better than hedges which have been well tended, so that large overgrown hawthorns were by far the most preferred. Hedges in which the undergrowth was either sparse or tightly managed held 2.5–10.0 pairs per km and were used principally by Yellowhammers and successively less by Song Thrush, Blackbird, Dunnock and Chaffinch, the last hardly using such vegetation at all. Overgrown hedges with outgrowths, on the other hand, held 15–37 pairs per km and were used by all five species. The Hertfordshire study of Fig. 7.2 and the East Anglian study of Arnold (1983) both support these findings in that wider hedges held more birds, with this effect being independent of the plant diversity effect.

Osborne (1982a) looked at whether the use birds made of individual hedges could be predicted from the structural features of each hedge. He plotted the positions of six common species in the course of nine CBC-type surveys and subjected the data to various statistical procedures. The statistical analyses were complemented by repetition of the survey in the following year, when many hedgerow elms had been felled (thus providing an experimental test of the statistical predictions from the first year's survey). Table 7.3. summarises the influence of each habitat factor on how many species and how many birds were present in each hedge, together with the number of species individually correlated with that factor. Nearly three-quarters of the variation in the number of species present was predictable, given the basal area of the hedgerow, the diversity of trees present in the hedge, the amount of scrub in the immediate surroundings, and the number of dead trees of 10–20 cm diameter in the hedge. The density of breeding pairs was also related to the presence of scrub and of dead trees but the nature of the surrounding fields had an even stronger effect, with fewer breeding pairs the greater the area of open fields about the hedge, and vice versa. The results largely agree with the more subjective assessments of Williamson (1971) on the same site: mature hedges with large trees (dead and alive) of a variety of species contained more species and individuals than did low-trimmed hedges, and scrub was a significant source habitat for hedgerow birds.

Cross-correlation amongst variables can make it difficult to interpret the results of statistical analyses. However, such problems were rather few both in our own study and in Osborne's study: hedges with numerous large live trees tended also to be close to rivers; hedges tended either to have many herb and shrub species and many small live trees or to be poor in all of these features; hedge area and ditch area tended to vary in parallel; and in hedges rich in herb species each species tended to be numerous. That

Table 7.3. *The influence of hedge structure and environs on bird abundance on a Dorset dairy farm*

	N of species correlated[a]	Correlation with species total[b]	Correlation with bird density[c]
Hedgerow features			
N of woody shrubs 0–10 cm DBH[d]	6	0.52	0.60
N of trees	5	0.55	0.49
Tree species diversity	5	0.52	0.38
N of live trees 10–20 cm DBH[d]	5	0.47	0.39
N of dead trees 10–20 cm DBH[d]	4	0.31	0.44
N of herbs	3	0.49	0.39
Tree growth form diversity	2	0.40	—
N of dead trees above 20 cm DBH[d]	2	—	0.39
Area of ditch bank	1	0.40	—
Live trees above 20 cm DBH[d]	0	0.40	—
Environs features[e]			
Scrub	2	0.44	0.58
Hedges	3	0.40	0.43
Field	0	0.31	0.30
Farmsteads	1	—	—
Woodland	0	—	—
Riparian habitat	0	—	—

[a] Number out of six common species (Blackbird, Robin, Blue Tit, Great Tit, Wren, Chaffinch) whose abundance in the hedge was statistically correlated with the habitat feature listed.
[b] Correlation between the number of common species present and the habitat feature listed. Dashes indicate correlation not significant.
[c] Individuals of common species present. Dashes indicate correlation not significant.
[d] Diameter at breast height.
[e] Environs defined as within 250 m of the hedge concerned.
Based on data in Osborne (1982a).

cross-correlations were otherwise absent means that most of the bird-habitat links found were mutually independent rather than due to some single general factor. Osborne in fact concluded that hedgerow communities were distinctive entities in their own right, rather than merely variants of the communities of woodland or urban habitats.

Osborne's study thus brings out for a single CBC plot the importance of the three hedge features found to be most variable between hedges by Helliwell (1975), namely hedgerow trees, shrub diversity, and amount of woodland in the area. Ecological evidence is available to suggest why birds should respond to each of the three variables. First, hedgerow trees provide additional foliage layers over which to forage (MacArthur, 1958). Second, a variety of tree or shrub species provides further structural diversity but also introduces a greater variety of invertebrate prey co-adapted to different plant species (Southwood, 1961; Bowden & Dean, 1977). Finally, the more woodland there is in the immediate area of each hedge, the greater the opportunity hedgerow birds may have to commute to richer food sources.

A second important conclusion emerges from this review. Individual species have quite different and distinct requirements in relation to hedge habitat. The densities of individual species are not always related to hedge density in a linear manner, as shown in Fig. 7.4 for Little Owl. Hence the linear correlation coefficients used in Table 6.2 in the previous chapter are at best a minimum estimate of the extent to which birds depend on hedges. Even so, that analysis showed that no fewer than 30 species responded to hedgerow density, with those more dependent on hedges than on any other habitat feature encompassing a diversity of taxonomic and guild classes, including two thrushes, three titmice, four finches, two corvids, and three migrant warbler species. These relationships vary markedly between

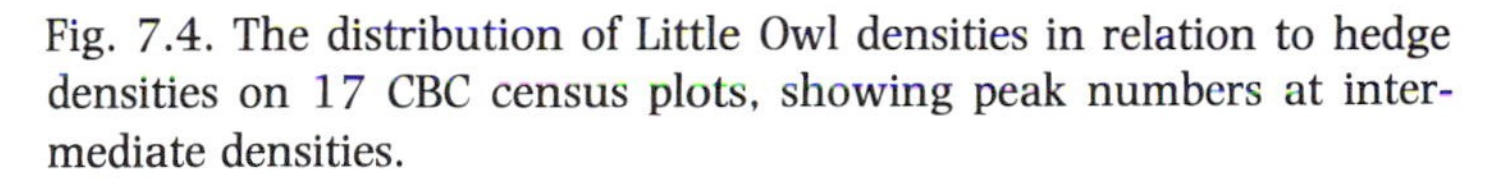

Fig. 7.4. The distribution of Little Owl densities in relation to hedge densities on 17 CBC census plots, showing peak numbers at intermediate densities.

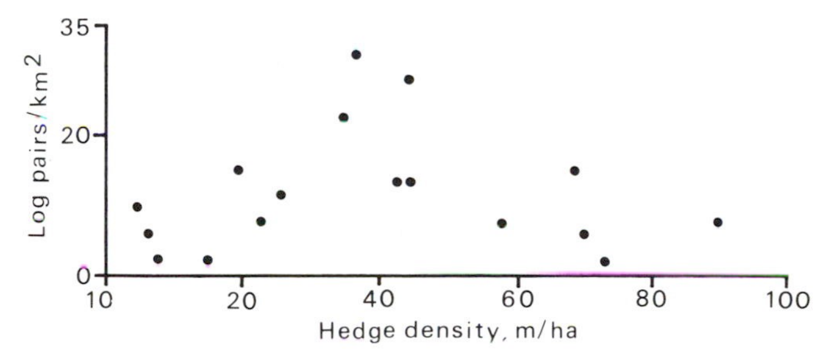

species, both where they are linear and where they are non-linear (Fig. 7.4 above). An immediate consequence is that no one management regime for a hedge will optimise the hedge for all bird species. Consequently, a diversity of hedge types will best ensure a diversity of bird species on a farm (Williamson, 1971; O'Connor, 1984a). The same point has been made both of hedgerow management and other habitat management in the studies by Arnold (1983) and Osborne (1982a).

Rates of hedgerow removal

Although parts of south-east and south-west England have long been hedged, many hedges in Britain have resulted from enclosure of open fields (Pollard *et al.*, 1974). Enclosure was first legalised in Britain in 1236 but in most Midlands counties less than 0.06 per cent of the available land was enclosed by the start of the sixteenth century (Pollard *et al.*, 1974). Legal enclosure of individual parishes began in 1603 and thereafter greatly increased, though with substantial variation between counties e.g. from 52 per cent in Northamptonshire through 40–50 per cent in most Midlands counties to as low as 4 per cent in Somerset and 0.3 per cent in Shropshire. However, the official records are misleading, for much enclosure of open fields in the late medieval and Tudor periods was done illegally. Later subdivision of already enclosed land to enable improved stock-farming or the new rotation farming (Chapter 1) also went unrecorded. Pollard *et al.* suggest that as much as 17 million acres must have been enclosed outside the scope of the various Awards and Acts of Enclosure. Hedges were thus added to the landscape throughout the late medieval to Victorian periods (by which time hedge removal was also well under way).

How much hedgerow is there in Britain? Locke (1962) conducted the earliest systematic survey of hedgerows in Britain, measuring hedgerow lengths in sample areas in eleven counties in 1954–57. This gave an estimate of 954000 miles (1.53 million km) of field boundary in Britain, some 65 per cent of it hedgerow, in 1954–57. Pollard *et al.* (1974), working from 1946–47 aerial photographs, estimated a total of about 500000 miles (804000 km) of hedgerow on improved farmland in England and Wales alone, based on an estimate of 13 miles of hedge per square mile (8.1 km/km²) of such farmland. Moore (1962) gives a figure of 616000 miles (991000 km) of hedge in Britain. However, such hedgerow is not distributed evenly throughout. Locke showed exceptional density of hedgerow in Devon and high densities in Yorkshire, Essex and Cumberland. His data show a marked north–south gradient in the proportion of hedgerow in field boundaries, with hedges common in the south and absent in the extreme north. This largely explains the corresponding gradient in the abundance of hedgerow birds from south to north shown in Chapter 2 (Fig. 2.7). Geographical variation in hedge abundance is thus mirrored in the distribution of the commoner hedgerow birds.

Hedge management and the history of hedge planting also varies regionally (Tozer & Taylor, 1979). In a study of farmland warblers in south-east Scotland S.R. da Prato (unpublished report) found that many of the hedges in this region were markedly poor for birds and suggests that two factors were responsible. First, Enclosure and its associated hedge planting was much later in south-east Scotland than in England. Second, hedge-laying was largely unknown as a hedge management practice, with the hedges instead kept closely clipped and rarely more than 2 m high. As elsewhere in these conditions, the presence of trees attracted more birds.

Since the war there has been a substantial disappearance of hedgerows from the English landscape under the impetus of agricultural modernisation. Pollard *et al.* (1974) give the following figures for hedge loss (miles per annum, with km per annum in parentheses) estimated from aerial photographs:

1946–1954	800 (1287)
1954–1962	2400 (3862)
1962–1966	3500 (5633)
1966–1970	2000 (3219)

Williamson (1967) estimated an annual loss of hedgerow of 1.1 yards per acre (2.5 m/ha) between 1963 and 1966, on 29 English and Welsh farms. This implies an annual loss of about 2.1 per cent of the 12 km of hedge and woodland edge that he estimated to be present per km^2 of such farmland. This figure overestimates both rate of loss and density of hedges since many CBC plot boundaries are of hedge and therefore shared with the adjacent area of farmland.

These data indicate an acceleration of hedge removal until the mid-1960s, with a subsequent slowing of the loss rate. But one problem in using aerial photographs in this way is that hedges cut to the ground but left with intact root stocks subsequently regenerate and have thus been lost only temporarily. Yet on aerial photographs they can appear as a total removal, depending on when photographed. Hence confirmatory evidence from other sources not subject to such a bias is required. One such source is the collection of habitat maps submitted to the BTO by CBC observers working on farmland. These maps show the gross features of each census plot, including the distributions of hedgerows, and are updated by the observer from time to time, By comparing the distribution of hedges on the various maps for each farm it is possible to assess the rate of change in hedgerow density over the years. Fig. 7.5 shows that annual hedge loss was about 1 m/ha through the 1960s but has gradually decreased. Our estimates are thus of the same order as those of Pollard *et al.* for the early 1960s and also extend to 1980. In Norfolk Baird & Tarrant (1973) estimated an annual loss rate of 1.4 m/ha between 1946 and 1970, though probably very much higher in the late 1960s.

Because hedgerow density is not constant over the whole of Britain, the impact of a given rate of hedge removal varies from place to place. This variation to a large extent reflects regional differences in agriculture in Britain. In the south-west dairying and stock rearing are still economically important and hedges serve to confine animals as required. But in cereal-growing areas hedges are not only useless to a farmer, they are seen positively as a nuisance, to be removed wherever possible. Pollard *et al.* (1974) found that between 1967 and 1969 grassland areas lost only 8.4 per cent of their 43.3 yards per acre (97.8 m/ha) whilst arable areas lost over 23 per cent of their 32.9 yards per acre (74.3 m/ha). Even so, there is a tendency on cereal land for hedge-removal rates to be correlated with the amount of hedgerow originally present, as found in Norfolk by Baird & Tarrant (1973). Hence although nationally hedge removal has been seen as particularly marked in the arable areas of East Anglia and as least in the south-west, the reality is that the highest rates have occurred where stock rearing has given way to intensive cereal cultivation (Fig. 7.6), i.e. where much hedgerow was present originally but where it now has least

Fig. 7.5. Rates of hedge loss recorded on CBC habitat maps in different years. Data extracted by BTO staff (G. Griffiths, R. J. Fuller, K. Taylor).

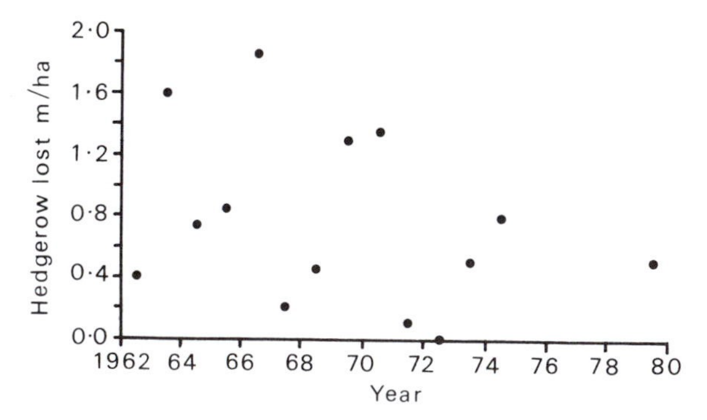

value. In this context it is worth noting that eighteenth and nineteenth century enclosures and their associated hedge creation had their greatest impact in the Midlands (Baird & Tarrant, 1973). The three reasons behind this large-scale removal of hedgerow in areas of cereal cultivation are, in

Fig. 7.6. Geographical variation in the annual rates of hedge loss in England and Wales, 1964–1976. Each point refers to the average of all data available for that county. Data as for Fig. 7.5.

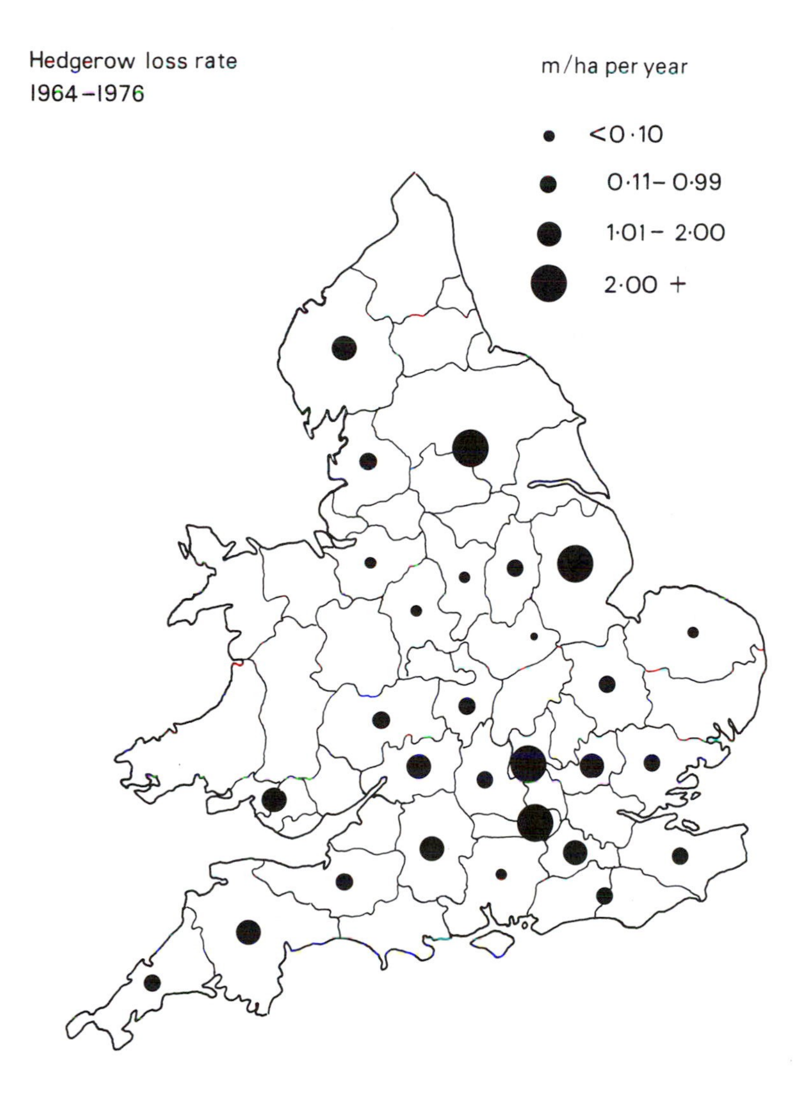

order of importance, the economics of machine use, the presence of reservoirs of cereal pests and disease, and the direct loss of land.

Modern farm machinery is large and powerful (Chapter 5) and most efficient when travelling in straight lines at its design speed. Anything that reduces the time spent under such conditions therefore reduces the average efficiency of the agricultural operation. Sturrock & Cathie (1980) looked at the main factors involved. These are the time spent turning (when the machine is idle), the time spent separately cultivating headlands, the time lost in moving from field to field, and the higher average speed possible in large fields. The relative economic importance of each of these factors varies with field size but a doubling of the average field size (6 ha) recorded in the CBC plots used in our farm habitat study would save about 17 per cent in working time. Sturrock & Cathie (1980) showed that a doubling of field size from 10 ha to 20 ha on a 200 ha farm yields a return of about 30 per cent per annum on the investment in hedge removal.

Some hedges have been removed as being reservoirs of crop pests and diseases but the case for this is unconvincing. For example, in cereal farming the major problems of aphids and aphid-borne diseases have emerged only recently, after the major reduction in hedgerows, and appear more related to the great increase in the area of cereals and to the effects of modern cereal management on populations of predatory insects (Potts, 1984a; Sotherton, 1984, 1985; Chapter 9). Nor are many of the economically harmful weeds of tilled land common hedgerow plants: rather, they are specifically adapted to the annual patterns of cultivation. The main exceptions are grasses but this problem is now increasingly dealt with by cultivating a narrow strip between the crop and the hedge to prevent such weeds spreading into the crops. Whilst the strip cultivation technique adds a little to costs it is simple and efficient and may be necessary whatever the nature of the field boundaries, e.g. even where hedges have been replaced by fences.

Several authors have estimated the amount of land lost to hedges that can be brought into cultivation on hedge removal. Baird & Tarrant (1973) quote a figure of 1 ha per 0.88 km of 2-m-wide hedge recovered. In practice one limit to the recovery of this land is the presence of ditches. Unless these are very deep they are not difficult to fill in but where they exist they are a necessary component of the drainage system and must be replaced by drains, adding to the cost of clearance. Even so, the returns indicated by Sturrock & Cathie (1980) suggest that banks and ditches are no real bar to reclamation of this land. Some ecologists have suggested, though, that a psychological element is involved in this issue, in that some farmers remove hedges because it is fashionable to do so (Terrasson & Tendron, 1981).

Several factors limit hedgerow removal and account for the decrease in removal rate since Williamson's (1967) estimate. Had it continued at that rate, all hedges in Britain would be cleared by the year 2010. First, clearance of hedgerow is expensive and so is most likely where there is a net gain in efficiency. Second, mechanised maintenance probably also reduces the cost of keeping hedges. Third, personal inclination against living on a prairie farm may be influential. Fourth, in some cases erosion has followed the removal of hedgerows and crop yields may also be depressed due to greater exposure. Fifth, in the west and north of England grassland and stock continues to be important and hedgerows useful: in France a 20 per cent difference in dairy yields between hedged and unhedged areas has been attributed to the shelter available in the former (Terrasson & Tendron, 1981). Finally, internal field boundaries are more vulnerable that are farm boundaries, and hedgerow removal is less likely where tracks, roads and ditches already exist along a hedgerow. We estimated such boundaries to constitute about one-third of the total hedgerow on CBC farmland. Pollard *et al.* (1974) found that hedgerow densities in Huntingdon settled over time at a level of about 7 yards per acre (15.8 m/ha), equivalent to a bounded (hedged) area of about 640 acres (259 ha) or about the average size of farms in their area. On the other hand, Baird & Tarrant (1973) estimated that hedge removal in Norfolk typically ceases at a level of 104 m/ha, considerably higher than the estimate of Pollard *et al.*

The impact of Dutch elm disease

A special case of hedgerow destruction in Britain has been the loss of elm trees (*Ulmus* sp.) to Dutch elm disease since 1969. The disease is caused by a fungus *Ceratocystis ulmi* whose spores are spread from tree to tree by bark beetles of the genus *Scolytus* and by transmission along the elm root systems that link many trees together. As the disease spread many trees died and others were felled in attempts to employ a sanitation cordon to halt the disease. Elkington (1978) estimated for selected counties that of some 17.1 million elms originally present, some 10.6 million were dead or dying and some 5 million had been felled since 1969. Osborne (1982a,b, 1983) has examined the effects on bird populations, which might be expected to appear through three major routes: reduced food availability; reduction in suitability and supply of nest sites; and reduced abundance of suitable song posts.

Most effects of elm loss on food availability are likely to appear only in the long term. Elm seeds are an important component of the diet of finches, forming as much as 20 per cent of the diet of nestling Greenfinches in May and June and forming a significant component of the diets of Bullfinch, Goldfinch and Linnet (Newton, 1967). Canopy loss to disease and felling could also affect species such as tits that take buds in winter and early spring (Gibb, 1954) or that take invertebrates from the foliage during the breeding season, e.g. Chaffinch, Tree Sparrow, Willow Warbler and Chiffchaff, and Goldcrest (Osborne, 1982a). Canopy loss also promotes the development of ground vegetation previously inhibited by shading, thus removing the bare ground over which the hedge-bottom Dunnock and Robin feed. In Dorset Nuthatches and woodpeckers appeared to benefit from the abundant food on dead elms but this effect must be only temporary since dead elms lose their bark after a couple of years, thereby removing the source of beetle larvae.

The effects of Dutch elm disease on nest sites have already been observed (Osborne, 1982a,b, 1983). Prior to the spread of the disease, elm trees were a particularly sizeable component of the nesting habitat on farmland in Britain. First, large tree cavities are rather scarce on farmland and elm trees are a valuable source of such nest sites for Kestrel, Stock Dove, Barn Owl, Little Owl, Tawny Owl and Jackdaw. Second, elms are frequently the most mature of hedgerow trees, thus providing for species such as Treecreeper and Tree Sparrow that depend on mature timber in various ways. Finally, elms are also the tallest of the hedgerow trees, thus accommodating species that prefer high nest sites, such as Rooks. Forestry Commission statistics indicate that some 16 per cent of the non-woodland elms in southern England were felled between the periods 1972–74 and 1975–78, so that a corresponding decline in elm use by birds would be expected. Declines greater than this must be due in part to diseased elms being less attractive for birds than were healthy trees, but smaller decreases could be due either to felling or to avoidance of diseased trees. Osborne found that both elm death and felling were contributing factors behind decreased elm use by Kestrels, Stock Doves, Tawny Owls and Barn Owls. Of these, Barn Owls were particularly badly affected (Burton & Osborne, 1980) and were, in general, unable to find alternative nest sites. The other species suffered in this way only where the clearance of dead elms was particularly extensive.

The extent to which the loss of tall song posts for territorial defence was limiting as a result of Dutch elm disease was difficult to establish (Osborne, 1982a). Different species were affected variously by the removal of dead or still living trees: Blackbird, Wren and Blue Tit were affected by the extent of losses of both live and dead trees but Great Tits were affected only by the loss of live trees. Only the Chaffinch was affected by the loss of structural diversity following the disease. As before, long-term monitoring will be needed if population changes are to be detected.

Overall, elm death and felling led to a loss of species, redistribution of territories and increases in territory size. Few species could be shown conclusively to be affected but Robins, Dunnocks and Goldcrests, and possibly Willow Warblers and Chiffchaffs, declined where the disease was rife and a number of other species (including Wren, Chaffinch, Greenfinch, Yellowhammer and Tree Sparrow) showed short-term decreases on CBC plots affected by the disease.

How important are hedges?

It is difficult to generalise about the consequences of hedge removal. Bird densities do not reflect habitat requirements wherever territorial behaviour and habitat quality interact (for example, when hedges are used only as overspill habitat when woods are fully occupied or vice versa) and this is commoner than is generally realised (van Horne, 1983; O'Connor, 1984a, 1986). This point is particularly relevant to the issue of hedgerow conservation since if hedgerows are regarded as suboptimal habitat occupied only as overflow from preferred woodland habitat, they might

Table 7.4. *Rates and speed of replacement of territorial birds of various species following experimental removal of the original territory holders*

	Percentage of population replaced	Days elapsed to first replacements	Days elapsed to new stable population
Blackbird	57	1–2	17
Dunnock	54	1–2	60+
Song Thrush	60	6	60+
Wren	100[a]	6	30+
Robin	100[b]	1	6
Great Tit	150[b,c]	6	24
Yellowhammer	75[b]	1	6
Chaffinch	0	—	—

[a] One captured bird escaped and could therefore have reappeared as part of the 'replacement' colonists.
[b] Includes some cases of neighbours expanding their territories into the vacant ground.
[c] More birds settled after removal than were present before.
Data from Edwards (1977).

therefore be of little consequence for the maintenance of populations (Pollard *et al.*, 1974; Murton & Westwood, 1974; Bull *et al.*, 1976). If hedgerow territories are not always filled, then birds may be able to move their territories in response to habitat manipulations such as hedge removal and might be able to maintain their numbers on the farm concerned. This has been recorded with Yellowhammers (Morgan & O'Connor, 1980), Robin and Dunnocks (O'Connor, 1984a) and Blackcaps (Osborne, 1982a) but the extent to which hedgerow quality is an important factor in permitting such adjustments is largely unknown.

For at least some species hedgerow does constitute a limiting resource, where all the available farmland territories are filled to saturation, and some birds are unable to acquire territories there at all. Table 7.4 provides evidence as to the presence of 'floaters' in several species common on farmland, including thrushes, Wren, Dunnock and Yellowhammer. Here the initial hedgerow territories were mapped and the owners were then trapped and removed to aviaries. Continued censusing showed, however, that new birds appeared and took possession of these territories. The inference is that these birds had been searching for suitable nesting habitat on the farm but were excluded by the territorial behaviour of the birds originally resident there. Note the variation between species as to both the extent and the speed of replacement, indicating different levels of population pressure within hedges.

Edwards' experiment was conducted in a cereal area in Hampshire where hedges were locally in short supply. In such an environment hedgerow territories tend to be very elongated, for hedgerow species are averse to moving long distances across open fields; this must limit local bird densities. Were additional hedgerow present, additional linear territories could be accommodated, allowing increased bird density. At very high densities of hedge, however, territories are no longer linear but may straddle whole fields, thereby reducing the number of birds the area can support, i.e. one bird has within its territory more hedge than it needs but the use of this surplus is denied to other birds. Of particular relevance here is that some species, e.g. Wren (Williamson 1971), Chiffchaff and Chaffinch (Osborne, 1982a), locate their territories at the intersections of hedges more often than expected by chance. Fig. 7.4 above shows that this is not merely theoretical speculation: instead, the highest bird densities of individual species may occur at some intermediate level of hedge density. Moreover, the optimum density differs between species, so that what is ideal for one species is suboptimal for another. This also means that the effect of hedge removal on bird populations depends on the initial and final densities of the hedges and on the species involved.

The various findings discussed in this chapter clearly necessitate a re-evaluation of the importance of hedges for bird populations. At the heart of the 'hedges are suboptimal' view of Murton & Westwood (1974) and Pollard *et al.* (1974) is the idea that hedges are an overflow or 'buffer' habitat with respect to preferred woodland habitat (Kluijver & Tinbergen, 1953). The strongest evidence in favour of this comes from Krebs' (1971) removal experiment with Great Tits, in which vacant territories in woodland were filled at the expense of hedgerow territories. The major weakness with this evidence is that different results were obtained in two subsequent repetitions of the experiment. In one case the vacancies were not filled at all (Webber, 1975), in the other they were filled but apparently by previously non-breeding individuals already present in the woodland rather than by hedgerow territory-holders (Krebs, 1977). The Woodpigeon removal experiment of Murton & Westwood (1974), however, provides circumstantial evidence for woodland preferences over hedgerow in this species. On the other hand, Edwards' (1977) removal experiment shows that hedgerow was filled to saturation for several common hedgerow species, so that hedge destruction would reduce the total carrying capacity of farmland for these species.

Area for area, hedges actually support more birds than does woodland. In one sense this finding is irrelevant to practical conservation, first because hedges are necessarily associated with areas of crop that do not support birds, and second, because increasing the local density of hedges above some optimum actually reduces bird density (Fig. 7.4 above). In another sense, the finding argues that hedges are good quality habitat, for the concentration of birds in woodland is itself partly an effect of habitat edge (e.g. Williamson, 1970) and this is less pronounced the larger the size of wood considered. In fact, the Moore & Hooper (1975) regression analysis of number of bird species on woodland area for each of 433 British woods gives an expected value of three species in a one-acre (0.405 ha) wood such as might be found on farmland, whilst the equivalent analysis for hedges gives an expected 6.6 species in a one-acre hedge (Osborne, 1982a). Given tight constraints on the amount of land – and therefore size of wood that might be devoted to wildlife on farmland – hedges emerge as a better use of the land!

Pollard *et al.* (1974) thought that hedgerow populations must contribute rather little to the maintenance of population levels in farmland species.

Such contribution is the product of density multiplied by breeding recruitment. Breeding success on farmlands is by no means so poor as to offset the density advantages of hedges just discussed (Chapter 3). It seems likely that the principal aims of nature conservation bodies in Britain will continue to lie with the protection and maintenance of semi-natural habitats (Ratcliffe, 1977). In addition, the residual value to farmers of small copses and spinneys for firewood and small timber, coverts in fox country, and so on, make it probable that small patches of woodland will inevitably persist on farmland. In such circumstances the evidence provided above argues that further field corner plantings are not the ideal target for wildlife management advice, such as might be offered by FWAG or ADAS advisers. Instead, the same amount of land devoted to hedgerows managed optimally and located to link the woodland patches might do much to maintain the matrix of habitat that characterises bird-rich countryside in Britain. Whilst preventing the spatial isolation of existing woodland 'islands' it could provide habitat intrinsically capable of supporting at least as many birds as the equivalent area of woodland. The optimisation of the layout and management of such hedge networks in the face of the operational constraints of modern farms has gone virtually unstudied but our data restore this question to the prominence it has lacked through the past decade.

Summary

Hedgerow bird densities are greatest where trees are present in the hedge, where shrub diversity is high, and where nearby woodland or scrub is abundant. The nature of hedgerow management is critical, for large overgrown hawthorns are particularly rich in birds. Individual species differ, however, in what maximises their density in a given hedge, so that there is no one optimal regime for all birds. Rates of hedge removal have slowed since the 1950s and have been highest where former grassland has converted to arable, particularly cereals. Dutch elm disease has adversely affected a number of species, largely through the loss of hedgerow trees. Assessment of the value of hedges for conservation requires taking into account the non-linear relationships between bird density and hedge density, as well as evidence that some hedgerow is fully saturated with breeding pairs.

8: The impact of farming practice on birds

Murton & Westwood (1974) distinguish two ways in which agricultural development can affect bird populations. The first is by engulfing and totally destroying particular habitats within farmland, such as woodlands and lowland heaths (Nature Conservancy Council, 1977; Moore, 1980). The second is by agricultural practice modifying the nature of the surviving habitats, thereby altering the niches they offer in ways that make them less (or perhaps occasionally more) attractive to birds. These latter changes in management techniques – within which we include choice of crops – are the subject of the present chapter. Table 8.1 summarises the main components of current agricultural practice and how they might be expected *a priori* to influence bird populations.

Since 1939 the conversion of permanent pasture to arable production has constituted a major loss of habitat within the farmland environment. Some of this loss has been replaced at the expense of rough pasture which has been improved and brought into greater agricultural production (as permanent pasture). These changes have not been the mere replacement of one existing habitat by another, however, for the changes in management that have occurred both in arable crops and in grassland have been as important. More intensive grassland management may lead to hedge removal and consequent habitat loss, for the highest stocking rates necessitate strict rotation of grazing over a series of paddocks, and this tends to favour replacement of hedges by wire and post enclosures. The impact of changed management on the farming year rather than on habitat has, however, been the dominant factor affecting birds, with arable fields nowadays affording birds a very different timetable of feeding and nesting opportunities than they had under traditional farming regimes.

The importance of management changes is supported by the experiences of many Common Birds Census (CBC) workers. Data gathered for the CBC includes information about management as well as about habitat changes and we asked the long-term CBC participants to comment on the management changes they had observed which had a significant impact on the bird populations of their plots. As with the habitat changes reported in Table 6.1, the results of this survey are probably biased but are still revealing. A summary of the changes thus identified is listed in Table 8.2. Of these, four – drainage, loss of permanent grass, increased tillage, and changes in crops grown – might be regarded as structural changes rather than as management changes and arguably could have been discussed within Table 6.1. Nevertheless, at least 97 (81 per cent) of the changes

Table 8.1. *Summary of the main changes in agricultural management and their implications for birds*

Change	Effect
Grassland–arable balance	Less pasture available for feeding Loss of breeding sites for ground-nesters
Loss of rotations	Reduced diversity of feeding opportunities
Hay–silage switch	Mowing in May–June, not June–July
Summer fallow loss	Loss of late sowing and tilling
Combine harvesting	Later, more rapid harvesting Less waste grain
Use of autumn-sown cereals	Earlier stubble clearing Loss of spring cultivations
New crops e.g. oilseed rape	New food sources
Methods of over-wintering cattle	Clumping of winter food sources
Chemical use	Fewer weed seeds and fewer insects Reduced breeding success

listed here are purely management changes. Management of stock (including both feeding systems and stocking rates) and disturbance were noted as the most frequent factors. Changes in the management of crops, in the timing of tillage and in methods of harvesting grass were each cited more frequently than the use of pesticides. Herbicides were the class of pesticide most often mentioned in this context. The most surprising feature of these comments was the importance placed on disturbance, a category which derived both from increased frequency of some farm work and the proximity of urban development to certain plots. Unfortunately the CBC scheme does not record data on disturbance but the table otherwise summarises management changes that are general and open to analysis for their effects on birds. We discuss our major findings in the following pages.

Skylarks and crop rotations

Skylark populations provide an interesting example of how changes in management can affect bird numbers quite as much as habitat changes.

Table 8.2. *The incidence of management changes on CBC farmland plots*

Methods of feeding cattle	6 (3)[a]
Increased stocking rates	12 (1)[b]
Drainage	11
Pasture improvement	5
Loss of permanent grass	4
Increased tillage	4
Shift from hay to silage	8
Change of crops cultivated	6
Changing crop management	8
Shift from spring to winter cereals	8
Hedge management	11
Disturbance[c]	16 (1)[d]
Herbicides	7
Game management	4 (3)[e]
Others[f]	12 (1)

Figures in parentheses indicate cases where the change was thought to favour birds.
[a] Three cases where cattle were concentrated into open stockyards in winter.
[b] One case of increased stocking of sheep leading to more carrion.
[c] Including disturbance by farm operations, mainly by spraying (six cases).
[d] One of reduced disturbance.
[e] One case of *reduced* game management, leading to fewer non-game species.
[f] Includes two cases each of increased predation by crows and by mink, persecution, use of slurry, use of chemicals (other than herbicides), and loss of rabbit grazing or burrows.
Data are from a survey of CBC participants' experience (see text).

Examination of four counties – Essex, Berkshire, Lincolnshire and the East Riding/Humberside – where the area of permanent grass has halved over the last 20 years shows that this had virtually no impact on Skylarks. Yet the species has nevertheless declined sharply since 1975 both there and in many cereal-growing areas. This change seems to be connected to changes in the area of leys, particularly of young leys.

Within individual farms Skylarks are often found to prefer just one of the several crops in common rotations and the changes in cropping usually cause the birds to rearrange themselves locally rather than to move longer distances. On the Westmorland farm studied by Robson & Williamson (1972) grass was always preferred to tilled land and roots were usually avoided altogether. On Oakhurst Farm in a six-year rotation of cereals, roots and clover, clover was preferred but distribution was otherwise fairly uniform. But when this changed to a seven-year rotation of five years of winter cereals, one year of spring cereals and one year of clover, the Skylarks increasingly concentrated in clover followed by spring cereals. Permanent grass was largely avoided in both periods. Finally, on a Hertfordshire CBC plot, distribution was very even until lucerne – grown as a four- or five-year ley for silage – was introduced and tillage crops

Fig. 8.1. Variation in Skylark numbers with age of leys on a Hertfordshire farm. A, B, numbers on two individual fields from year to year; C, average density over leys of the stated age.

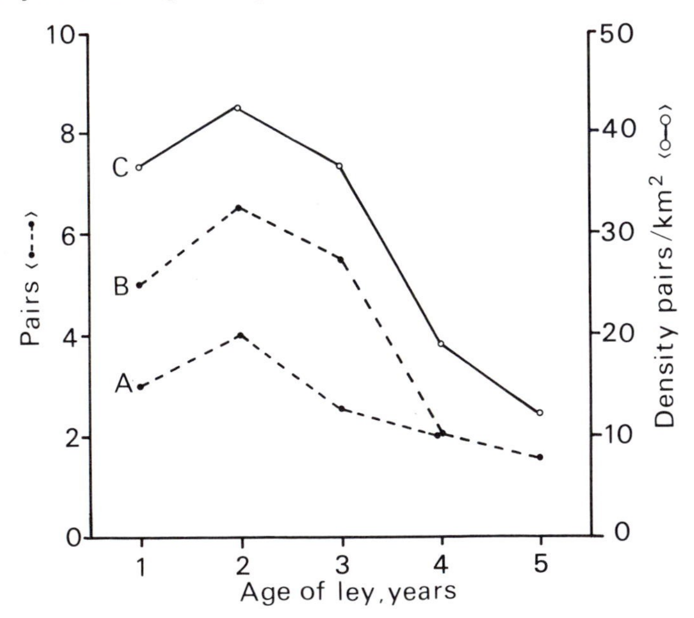

largely dropped; the birds then showed a marked concentration into lucerne and grazed leys. These results show that, although Skylarks do prefer some crops over others, the preference manifest at any one time may depend on what is available on each farm rather than reflect some absolute predisposition of the birds.

On the Hertfordshire farm Skylarks selected young leys, usually shifting when these leys were more than three years old (Fig. 8.1) and this was true also of the Westmorland plot. In Sussex clover leys were one-year leys only and so moved annually, and the birds showed their preference for them by following these leys around the farm. The preference indicated on these farms suggests a positive link between Skylark numbers and the age of leys. Since 1975 the MAFF June Census has recorded the area of leys under five years old. Expressing this as a percentage of all grass (except rough grazing) and plotting the national CBC index for Skylark against it results in Fig. 8.2, showing a striking correlation. This association with young leys is probably also reflected in the greater breeding success in mown grass noted in Chapter 3 (Fig. 3.3). About half of all young leys are

Fig. 8.2. Skylark population levels in Britain (as measured by the CBC index) in relation to the percentage of young leys (under 5 years) in the grass acreage of England and Wales (rough grazings excluded). Data are for 1975–83 except 1982, when the population was depressed by the severe 1981–82 winter.

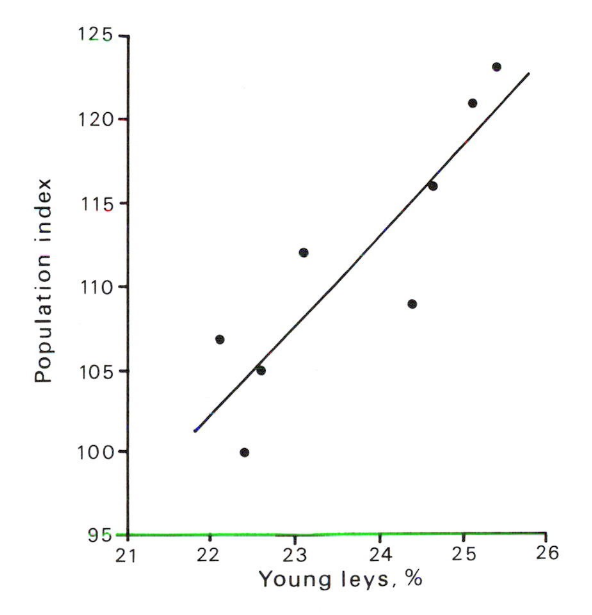

mown, as against only a quarter of permanent grass. Nationally, the acreage of young leys was rather stable in the mid-1960s but since 1975 leys have declined steadily in England and Wales, from 1 337 915 ha in 1975–77 to 1 131 553 in 1981–83, i.e. by just over 15 per cent. The overall decline in Skylark numbers noted above seems therefore to be related to this general decline in young leys.

These observations are based on territorial or singing birds, which may not be the best basis for examining the use any ground-nesting birds make of crops. In rotations territories may shift because either the nest or the feeding area has to follow the rotation. Lapwings, for example, tend to nest in one crop but take their chicks to another (Redfern, 1982) and both are necessary to it. Knight & Shepherd (1985) have shown that Skylarks may group nests in a favoured part of an area, whilst feeding elsewhere. Here it may be useful to compare Skylark and Corn Bunting. Both general CBC analysis and a detailed analysis of the Sussex and Hertfordshire farms considered above show that the numbers of Skylarks and Corn Buntings fluctuate very closely in parallel. This suggests that both species are affected by the same factors. But examination of the territory distribution on both farms shows significant differences in the selection of crops, although both species react to the rotation. CBC plotting of singing birds is therefore only a crude descriptor of the birds' responses to crops and crop management. It seems very likely, in fact, that partitioning of use amongst two or more crop types within common territory boundaries is a normal pattern with ground-nesting birds using crop habitats.

Lapwings and levels of management

The Lapwing is a species less sensitive than the Skylark in response to rotational changes but it is more sensitive to the intensity of management. The species has been the subject of two national BTO surveys, in 1937 (Nicholson, 1938–39) and in 1960–61 (Lister, 1964). The latter, in particular, provides a valuable baseline from which to measure the effects of the changes in agricultural land management that have taken place over the period of the CBC. Lister found evidence of regional variation in the fortunes of the Lapwing since the 1937 survey, i.e. over the course of the post-war agricultural revival and the major shift in the arable–pasture balance that accompanied it (p. 89). Survey returns from areas experiencing a marked return to cereal production, such as the East of England, showed a preponderance of declines since 1937, whilst the opposite was true of the more pastoral areas of the country. This trend has continued since Lister's survey (Fig. 8.3), with Lapwing densities in pastoral areas nearly quadrupling in a recovery from the cold winter of 1962–63. In cereal-dominated areas numbers have fallen, more or less continuously since 1963, to about a third of their 1962 level. These trends have occurred at a time when the arable–pasture balance was stabilising but when management changes were intensifying, and they partly underlie the major differences in the 1960–61 and 1982 distributions (Lister, 1964; Smith, 1983).

In tilled land Lister found a preference for nesting on spring cultivations and this preference is still present today. Galbraith (1985) demonstrates that, over extensive areas, Lapwings prefer nesting on spring-tilled land,

Fig. 8.3. Trends in Lapwing populations in areas of England and Wales dominated by cereals (top) and sheep rearing (bottom), 1962–83.

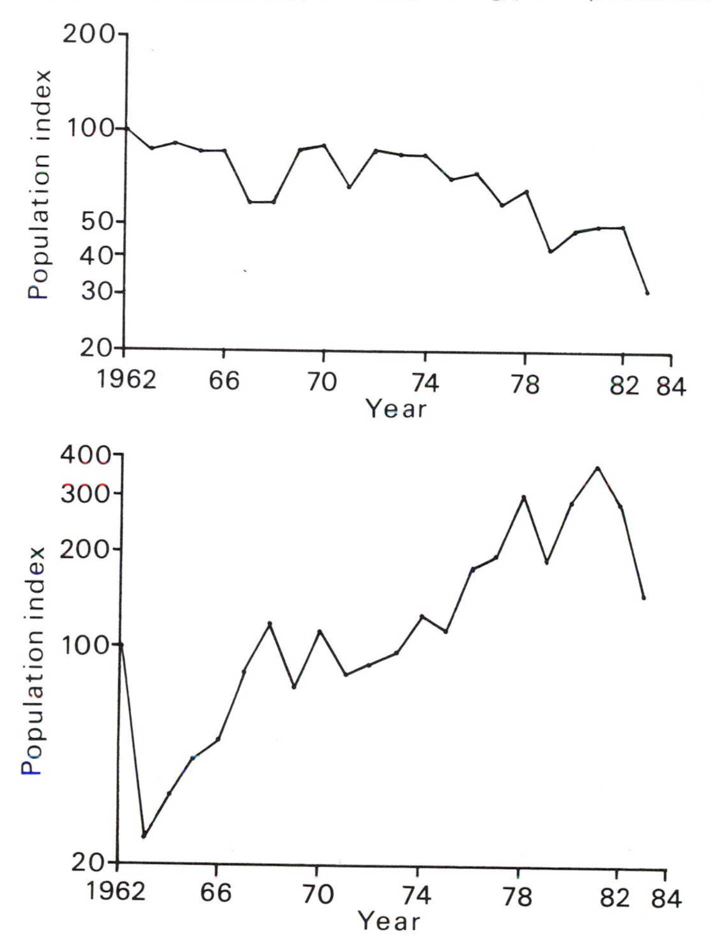

although local densities on grassland can be very high. Chicks are usually moved to grassland for feeding and this probably limits breeding distribution in some areas. These habitat shifts in the course of the nest cycle suggest that tilled land is better for egg production or incubation and that grassland is better for chick feeding and rearing. This may be the result of different seasonal changes in food availability in the two habitats (cf. Murton, 1971) or because nest camouflage is better in arable (cf. Klomp, 1953). The nest record cards for various counties respectively rich in tillage and rich in grassland may support the former explanation: over the period 1962–80 clutch sizes were about 4 per cent larger in tillage areas (average 3.77 against 3.62 in grass) and hatched more successfully (55 per cent against 42 per cent). Egg formation and date of laying in Lapwings has been shown to be directly linked with the availability of earthworms in spring (Hogstedt, 1974) and these might be expected to be more available (though probably not more abundant) in freshly worked soil. On sugarbeet, though, beetles are a more important food (R. Green, personal communication) and on the North Downs of Hampshire many birds nesting on arable commute to grassland to feed (T. P. Milsom, personal communication). Other explanations, such as seasonal changes in predation or in vegetation height, are possible. Lapwings protect their eggs by aggressive defence of the nest site, so egg success is sensitive to their ability to detect approaching predators. In heavily vegetated areas, therefore, Lapwings prefer to nest in the barer patches. The nest record analyses also show, as expected, that tilled land is poorer for chick rearing, perhaps because this openness also exposes the feeding chick to detection by predators. Over the period 1962–80 chick success rates in tillage and in grass have been fairly similar on average, at 35 per cent and 38 per cent respectively, but the situation has been dynamic. Over the last five years of the period the figures have averaged 25 per cent and 40 per cent respectively, i.e. tillage has worsened as a habitat in which to rear chicks. Grassland appears to have improved as chick-rearing habitat over the period, with chick success rising from 18 per cent in 1962–65 to 26 per cent in 1966–70, 44 per cent in 1971–75 and 40 per cent in 1976–80. In contrast, chick success in tillage areas had fallen to 25 per cent in 1976–80 but showed no systematic trend through the period. Brood success is very difficult to establish for precocial wader chicks, however, so these data are merely indicators of what may be happening.

In theory the decline in spring tillage could result in Lapwings moving to alternative habitats, towards greater use of autumn cereals, for example. Modern management changes have, however, also affected these other habitats and reduced their suitability for nesting. The switch from spring-sown to autumn-sown cereals, for example, has resulted in a 30 per cent reduction in Lapwing densities and in later laying and smaller clutches (O'Connor & Shrubb, 1986). This decline of Lapwing in cereal areas is probably related to changes in crop density and height. The main management changes in autumn cereals have been the introduction of tramlining techniques, the use of autumn herbicides, very early nitrogen dressings on wheat crops, and the use of foliar fungicides. These changes have been progressively introduced since about 1969 and the main impact has been to increase the density and evenness of plant stands in the early stages of the growing season. Lister showed that in the early 1960s Lapwings would nest in autumn-sown cereals up to 12 inches high, but by 1975 Glutz *et al.* (in Cramp & Simmons, 1982) found that dense crop stands were deserted at c. 3.5 inches high, although thinner stands were used up to about 6 inches. Glutz *et al.*'s data are for European populations but the comparison is valid since crop densities have risen across Europe. The likelihood that increased crop densities have pushed Lapwings out of cereal crops is also shown by individual case studies. For example, counts of nesting Lapwings at Oakhurst Farm over a 25-year period show that autumn cereals were used regularly until the early 1970s, when the management changes just described were introduced. These changes greatly increased plant densities in the early spring (from 221 plants/m^2 in 1978 to 280 in 1984, for example) and promoted a more even plant stand. Despite a significant increase in the area available, densities of nesting Lapwings in autumn cereals then fell from 6.5 pairs per km^2 in 1961–65 to 3.5 in 1975–80 and 1.5 in 1981–85, when nesting there was in fact irregular.

That these effects have been widespread was shown by a county-wide survey of Lapwings in Sussex (Shrubb, 1985b). Here the habitats visited included 47.6 km^2 of autumn cereals spread across the coastal plain, the Downs, river valleys, and the Weald. Of these, 23.4 km^2 were tramlined (and therefore probably intensively managed) and 24.2 km^2 were not. Of 34 pairs recorded in these crops, only 4 (12 per cent) were found in tramlined fields. Furthermore, less than 30 per cent of the autumn cereals visited in the Weald were tramlined and this crop was the principal nesting habitat there. Here poor drainage and very heavy soils often make autumn cultivations difficult, resulting in frequent bare or poor patches in fields, which Lapwings use. On the Selsey Peninsula, probably the most intensively farmed area of the county for cereals, the result of these management changes has been to concentrate virtually the whole nesting population of eight farms totalling 1500 ha into two areas of permanent grass totalling 160 ha, at an extraordinary density of 120 pairs per km². Similar displacements have also been recorded of Lapwings in Hampshire (T. P. Milsom, personal communication).

Waders, grassland management, and stocking rates

Intensive management in grassland is usually possible only where adequate field drainage is installed and almost invariably requires reseeding with high-yielding rye-grass varieties to make best use of nitrogenous fertilisers. It also involves high stocking rates, high fertiliser usage, a preference for silage rather than hay, and more prevalent multiple cutting of grass. The switch to silage from haymaking is almost entirely a post-war development and now affects up to 30 per cent of the grassland in the dairying areas but only 11 per cent in arable and livestock areas (Church & Leech, 1983). Whereas hay meadows are normally mown in June and July, silage is cut in May and June and up to three cuts may be taken. The earlier cutting affects the timing of fertiliser applications, and the greater frequency of multiple cutting separately affects the quantity of fertiliser applied. Nearly 2.5 times as much nitrogen is used on silage grass as on hay: 35 per cent of fields cut for hay received low nitrogen doses (under 50 kg/ha) and only 20 per cent high doses (more than 150 kg/ha); for silage grass the corresponding figures were 5 per cent and 66 per cent respectively. As a result, silage fields show a much denser and lusher plant stand from April. This must render silage grass useless as a chick-feeding area for Lapwings, for Lister noted as early as 1960–61 that silage grass in April was even then too high for Lapwing, the birds leaving dense crops of more than 6 inches in height (cf. the case with cereals above). Hay grass is thus inherently more likely to be favoured for chick feeding. The high nitrogen use early in the season also intensifies the green colour of grass fields, a factor which Klomp (1953) showed to influence site selection by Lapwings.

Ground-nesting birds are markedly affected by the intensification of grassland management. In Switzerland, for example, intensive grass held only half the species and a quarter of the density of birds on predominantly arable land (Luder, 1983). In another Swiss study, the numbers of birds breeding within a nature reserve increased as populations in the surrounding agricultural land decreased with the lowering of the watertable there (Fuchs, 1982). Ground-nesting birds are particularly vulnerable to any management alteration of the timing of farm activities. The decline of the Corncrake in Britain and Ireland, for example, has been attributed to the change in timing associated with the shift from haymaking to silage preparation (Cadbury, 1980a, 1984). Corncrakes lay in June, principally in meadows, and the chicks hatch some 15–18 days later, when they can be led to undisturbed marshes when the meadow is mown. Earlier silage cutting means that nests in silage still contain eggs and are therefore destroyed when the field is cut, thus reducing overall production of chicks.

Smith (1983) has described the distribution of Lapwing in grassland in lowland England and Wales and his results show a marked coincidence between Lapwing numbers and less intensive grassland management. In the mainly arable areas (as defined by Church & Leech, 1983), Lapwings averaged 37 pairs per 1000 ha of grassland. In contrast, in the more intensively managed grassland of the mixed arable/dairying regions and of the mainly dairying regions, the grassland supported only 24 pairs per 1000 ha of grass. This relationship is explored further in Fig. 8.4 which shows that Lapwing numbers in each MAFF region are reduced as the average stocking rates increase, even though the regional variation in stocking rates is only 23 per cent. This is not a large variation in terms of animals per acre but it is superimposed on historically very high absolute rates of stocking. These rates are now at their highest since the June Census began in 1866 and have doubled since the 1930s; they have increased by 36 per cent since as recently as the early 1960s.

Fig. 8.4. The density of nesting Lapwings in lowland grassland in each region of England and Wales in relation to average stocking rates. YL, Yorkshire–Lancashire; N, Northern England; SE, South-east England; E, Eastern England; SW, South-west England; EM, East Midlands; WM, West Midlands; WA, Wales.

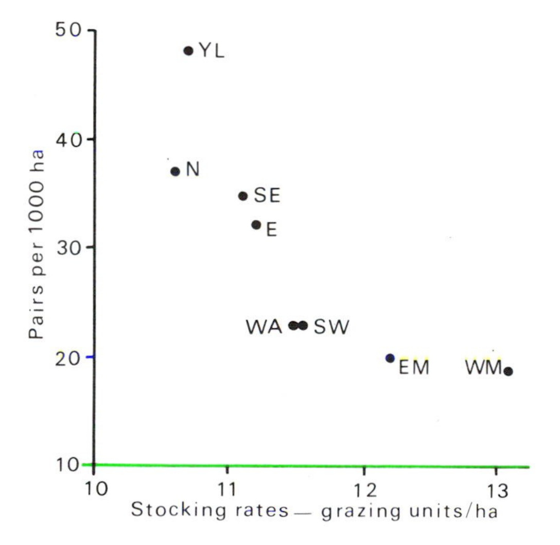

The evidence of this figure and of Fig. 8.2 above for Skylark suggests that ground-nesting birds in Britain may now be very sensitive to changes in grassland management, for both figures show sharp reactions to comparatively limited changes. In the case of leys this probably results from the way in which crop distribution is affected by declining crop area. Crops do not decline uniformly over the farmland area: after a certain point they tend to disappear altogether from some areas and to concentrate in others as specialisation continues. There are therefore limits to the extent of farmland modification that bird species will tolerate, so that steep population declines occur if the changes go beyond these limits.

The Snipe is a species particularly threatened by the widespread intensification of drainage. Pasture accounts for 52 per cent of all Snipe habitats recorded in the Nest Record Scheme and two-thirds of these records are from wet pasture (Mason & Macdonald, 1976). The sample may be biased towards pasture, though, since this habitat is more accessible than bogs or moorland and therefore more likely to be visited by BTO nest recorders. The structure of the sward is particularly critical to Snipe, with

Fig. 8.5. Modern grass fields, showing the effect of strip grazing and high stocking rates.

89 per cent of all nests in grass or rushes and 58 per cent of the grass sites expressly described as in tussocks. In pasture this structure is directly related to grazing pressure, which can affect Snipe both through habitat modification and through trampling losses. The first occurs because the sward becomes less tussocky as the number of animals present increases. In fact, close grazing combined with a summer mowing will slowly eliminate rushes without recourse to herbicides, even in quite damp pastures. The importance of a tussocky structure to Snipe is probably related to the need for nest concealment to foil predators, since predation was responsible for 56 per cent of the identified causes of failure in Mason & Macdonald's study. Trampling losses are the second result of higher stocking rates. In Britain Mason & Macdonald found that 25 per cent of nest losses among Snipe resulted from trampling, mainly by cattle. These two sources of nest losses under higher grazing pressure brought about by post-war increases in stocking rates may largely account for the decline in abundance of this species. Mason and Macdonald have shown that a large proportion of those nests which fail must be replaced each season if each pair of adults are to replace themselves and this indeed seems to be the case (R. Green, personal communication), but any increase in losses must then result in a net decrease in population size.

The problem of wader nest losses to trampling by stock (Fig. 8.5) has been well studied in the bird-rich meadows of Friesland (Netherlands) by Beintema and his colleagues (Beintema, 1982; Beintema *et al.*, 1982) and in eastern England by R. Green (personal communication). These studies show that Lapwings are less vulnerable than other waders to trampling damage by sheep or cattle. About 40 per cent of Lapwing nests are trampled at stocking rates of one cow per acre but Snipe lose 60 per cent and Redshank 72 per cent at this density of cattle. At double this stocking rate, i.e. near the upper limit to current levels, the three species respectively lose 60, 85 and 93 per cent of their nests (R. Green, personal communication). Beintema (1982) found that one day of grazing by 20 animals gives exactly the same losses as 20 days by one animal but the consequences for the breeding population differ because birds that lose clutches may lay again, as in the Snipe discussed above. Both Beintema and Green show that the earlier grazing of higher densities of cattle associated with pasture improvement prevents successful re-laying, and that Redshank and Snipe are worse affected than Lapwing. The earliest nesting pairs rear more young than late pairs do in such conditions and in the Netherlands this has resulted in a marked shift in the timing of breeding of meadow waders: Lapwing, Redshank, Snipe and other species now nest there one or two weeks earlier than they used to do at the beginning of the century (Beintema, Beintema-Hietbrink & Muskens 1985). Advancement of laying in this way must eventually be counteracted by other considerations, such as greater predation of nests in very short vegetation or difficulties in finding enough food to form eggs (Klomp, 1953; Perrins, 1970).

The loss of spring tillage

The seasonal pattern of farm work has to be accepted by every animal inhabiting farmland but the changes in the periods of tilling the soil associated with the introduction of winter cereals have reduced the availability of food supplies to birds, particularly in the breeding season. The majority of birds seen on farmland in spring are recorded on fresh cultivations; few birds are seen either on grassland, unless stock are present, or on autumn cereals. Fig. 8.6 shows the long-term consequences for Song Thrushes of reduced availability of spring tillage brought about since 1975 by a shift from predominantly spring-sown varieties of barley to predominantly autumn-sown ones. The only other notable change in these counties appears to be a decline in ley grass but, unlike the Skylark above, there is no clear correlation between the CBC Song Thrush index and ley grass. The data thus suggest that the Song Thrush population fell in response to the reduction in spring cultivations and the feeding opportunities these provide. One possibility is that Song Thrushes returning to the area from their winter habitats might reject the areas of autumn-sown barley as suboptimal habitat and move elsewhere in search of better habitat: the CBC would then detect this as fewer territorial birds in the area. A second possibility is that the birds do take up territories in the winter barley areas but breed there less successfully because of the poorer feeding

Fig. 8.6. Population trends in the Song Thrush in areas of England and Wales dominated by barley cultivation, 1962–83.

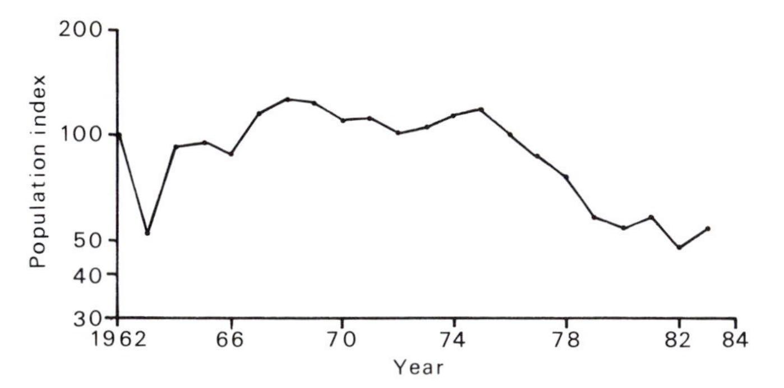

conditions experienced when the frequency of spring tillage is reduced. Our nest record data for the same counties and time period support both ideas: where 51 per cent of all eggs hatched successfully under the spring-sown barley regime, only 37 per cent did so with autumn-sown barley dominant. We then calculated what effect such a change in hatching success would have if no other change in breeding parameters or in survival took place

Fig. 8.7. Mistle Thrush populations on CBC plots in areas of England and Wales dominated by: (*a*) sheep rearing, (*b*) cereal management, and (*c*) barley management. Data in sheep rearing areas are sparse. Note the use of a logarithmic scale and the use of 1966 as the arbitrarily chosen reference year (index = 100).

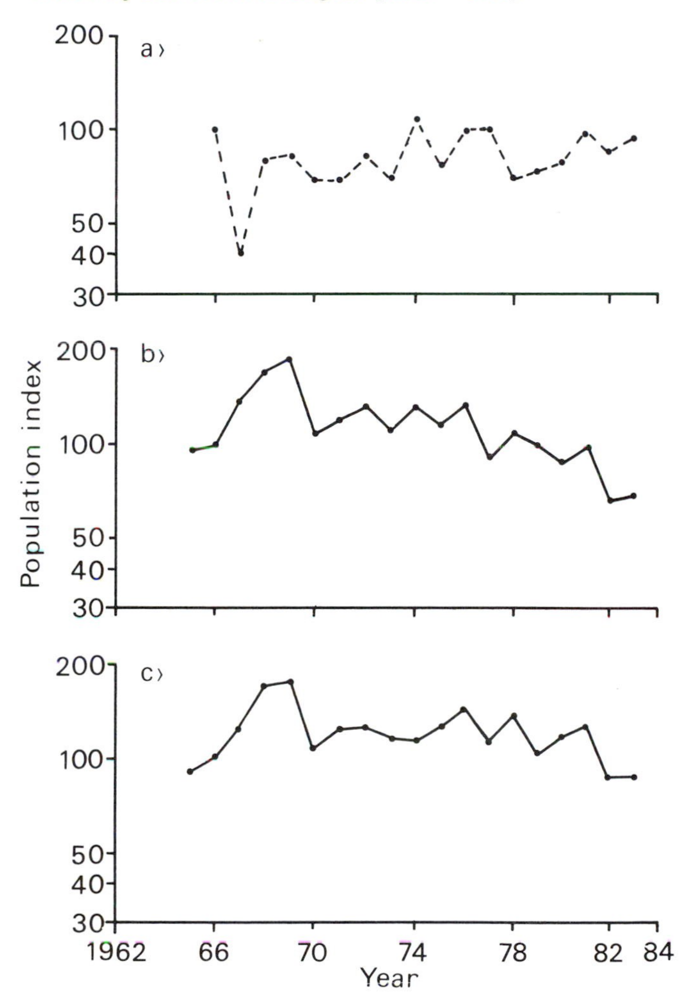

over the period. The calculations show that, if the local population were self-sustaining, a fall in hatching success of this order would result in a decrease in population level of the order of 27 per cent. This is only about half the decrease actually observed (Fig. 8.6) and suggests that some birds that would normally have bred in the affected areas now go elsewhere.

The Mistle Thrush is another ground-feeding species affected by the shift in timing of cereal crops. As it is the earliest of the thrushes to breed each year, it might even be expected to be the most severely affected where early spring foods are limiting (cf. Perrins, 1970). Fig. 8.7 shows, however, that its sensitivity to the loss of spring cultivations is not shown in any uniform trends in the different agricultural regimes considered by the CBC. The differences are largely due to the bird's readiness to forage widely, travelling up to a kilometre in search of food for its young. In regions dominated by pasture or by sheep, population trends have been at a more or less constant level from 1968 to 1983. In areas dominated by cereal and barley, however, a reversal of a recovery from the effects of the 1962–63 winter is apparent about 1970–71 (Fig. 8.7). In the cereal-dominated sample the start of this decline coincides with the increase of winter cereals, particularly wheat. In the barley-dominated sample the decline is less severe, as spring barley remains a widespread crop and provides a source of spring food accessible to a mobile species; this allows it to continue breeding in areas which would otherwise be unusable. Long-distance commuting to food sources must of course be more costly than local foraging but many Mistle Thrush habitats are saturated at current densities (O'Connor, 1986) and the birds probably have no alternative. The nest record samples specific to individual agricultural regimes have rather wide confidence intervals, probably associated with small sample sizes, but suggest that chick-rearing success in cereal areas may have decreased since the 1960s, from 39 per cent in 1962–65 to 28 per cent in 1976–80. The Starling provides a further example, for it continued to increase even in the face of pasture loss until the mid-1970s, but after the disappearance of spring tillage on the cultivated lands in these counties, numbers there quickly fell to low levels.

Breeding success in a number of other species has also been affected by the loss of spring tillage, amongst them Rook and Lapwing, even though these are additionally affected by other management changes (see above). In the case of Lapwing we have shown that laying is more often late and clutches smaller where spring tillage has been lost (O'Connor & Shrubb, 1986).

The expansion of cereal farming

Mechanisation and Rooks

The widespread mechanisation of cereal harvesting has had a marked impact on the ecology of the Rook and other species which favour a mixed grassland–arable regime (e.g. Brenchley, 1984). Grassland is especially important to the Rook for summer feeding. Although grazed permanent pasture is preferred, newly mown grass is hunted over for insects and the switch from haymaking to ensilage has brought this insect peak forward to earlier in the year. This advance would have been rather unimportant had traditional harvesting methods continued in any adjacent cereal fields, for grain was still available as an alternative food source from late July through to early September in the south (October in the north) (Fig. 8.8). In modern cereal farming, however, the use of combine harvesters means that grain is left to ripen on the stalk and is harvested more rapidly. Fewer stages, each of which used to involve waste, mean that less grain is lost and the low labour content of the operation subsequently allows stubble clearing to start earlier. Harvesting thus starts later and lasts a shorter period in total and results in less waste grain which is cleared more rapidly. Moreover, the use of summer fallows has declined with the advent of

Fig. 8.8. Pre-combine harvesting with self-binders left much waste to be exploited by Rooks.

powerful herbicides, thus removing a possible alternative source of food. This has been further aggravated by the widespread decline of spring tillage (p. 86). The overall result is that the period of insect availability has moved forward in time and the period of grain availability has moved back in time, thus widening the 'hunger gap' in which the bird is without an abundant food supply and in which most mortality occurs (Murton, 1971; Feare *et al.*, 1974).

The regional avifaunas and two national surveys provide much information on the pattern of change in Rook populations in recent years. Dobbs (1964) showed that Rooks in Nottinghamshire increased substantially between 1932 and 1958, when cereal growing was expanding, and that this increase was unrelated to changes in the acreages of grass or tilled land in the county. The increase was attributed to more land being under corn (Dobbs, 1975). A similar increase in Gloucestershire between 1930 and 1950 was followed by a decline through the 1960s and 1970s (Swaine, 1982). In Derbyshire a local decrease took place between 1944 and 1965–66 in the Sandstone Belt and Southern Gritstone regions but the species increased in the High Peak district (Lomas, 1968). Numbers increased between 1944 and 1958, then decreased thereafter, especially in grain-growing regions. In 1970 Lomas compared two regions near Ashbourne and found further decreases in grain areas but no change in numbers in limestone grass areas (Frost, 1978). In Kent the species declined by 68 per cent between 1949 and 1975 (Taylor *et al.*, 1984) and in Surrey the species was at a low ebb in the early 1960s, perhaps due to the contamination of food by toxic chemicals (Parr 1972). These varied population changes have been explained by Brenchley (1984) who showed that for Rooks the optimum habitat is one containing a mixture of tillage crops (cereals, roots and vegetables) and of grass (temporary and permanent but excluding rough grazing) in a ratio of 45:55. This is also apparent in some of the regional data. In Sussex, for example, Rooks are concentrated into areas with a good mix of arable and permanent grassland (Shrubb, 1979) whilst within the area covered by the West Midlands Bird Club (Staffordshire, Warwickshire and Worcestershire) densities were highest in river valleys supporting mixed farming with good numbers of hedgerow trees and small copses (Harrison *et al.*, 1982). The species therefore decreases if either the grass component or the cereal component of mixed farming alters away from the optimum ratio. We have examined regional variation in clutch size and breeding success of the Rook in this connection. In eastern England clutch size averaged 3.8 against 3.6 in southern and western England but nest success in the east was only 45 per cent against

75 per cent in the south and west. Thus overall productivity is greatest where mixed farming is more frequent.

Habitat changes have probably had a second, though not inconsiderable, influence on Rook numbers in Britain. An increasing loss of elms to Dutch elm disease has had a marked effect on the north Kent landscape and is thought to be responsible for the decline of rookeries there (Taylor *et al.*, 1984). Rookeries in elm trees were only half the size of those in oak (Dean, 1974) but in the West Midlands elm accounted for over half of all nesting trees (Harrison *et al.*, 1982). Fig. 8.9 shows how Rook densities may have declined recently in modern cereal areas, though as the CBC method was not intended as a census technique for Rooks there is much scatter in the data. Too few nest records are available for the Rook to trace the history of breeding success in these areas but our conclusions are reinforced by those of two national surveys of the species in 1975 and 1980 (Sage & Vernon 1978; Sage & Whittington, 1985). These showed that major decreases in rookery numbers since 1944–46 were concentrated into eastern England. Analysis of local surveys that have been conducted in the intervening period showed that much of the decline took place in the 1960s, though continuing at a slower rate into the 1970s. Similar declines took place in Scotland and Wales (Sage & Vernon, 1978).

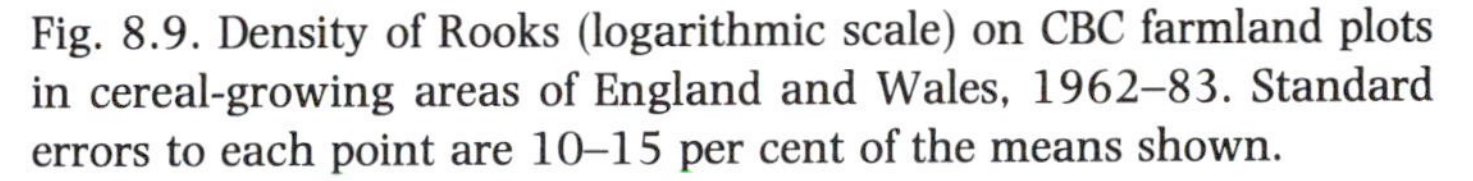

Fig. 8.9. Density of Rooks (logarithmic scale) on CBC farmland plots in cereal-growing areas of England and Wales, 1962–83. Standard errors to each point are 10–15 per cent of the means shown.

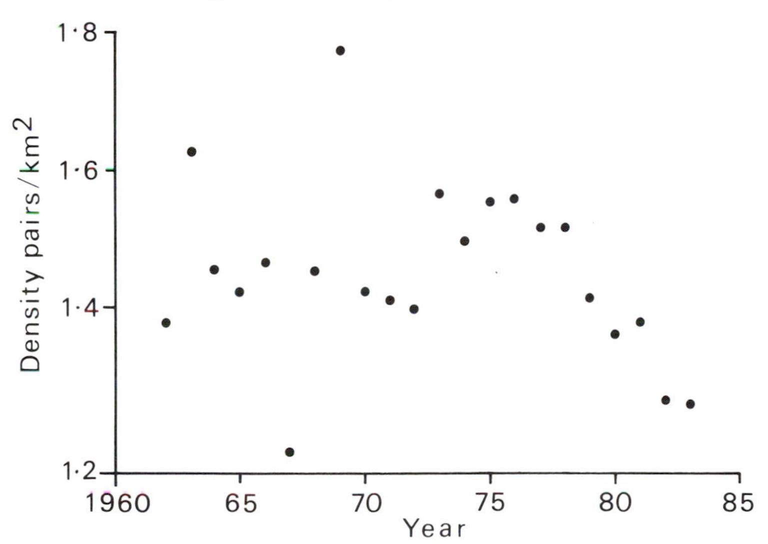

Fig. 8.10 suggests why Rooks in the more heavily tilled counties of England might be more vulnerable to changes in cereal practice than those in western counties. These results are from the 1944–46 Rook Investigation conducted by the BTO. They show that even then, before the further intensification of cereal cultivation in these counties, Rooks in the eastern counties were taking fewer animals and more cereals in their diet than birds in western counties. Hence subsequent changes in cereal-growing methods will have fallen more severely on Rooks in the east, a point evident in the most recent survey results.

Cereal management and Woodpigeon breeding seasons

Woodpigeons have proved particularly sensitive to the changes in timing involved in modern cereal management. Woodpigeons have a long breeding

Fig. 8.10. Comparison of the volume of major foods in the diet of Rooks in two areas, by months. The Southeast area consisted of Kent, Sussex, Surrey, Middlesex, Essex and Hertfordshire. The Southwest consisted of Cornwall, Devon, Somerset, Gloucestershire and South Wales. The vertical scale is marked at intervals of 10 per cent of total volume. After Holyoak (1972).

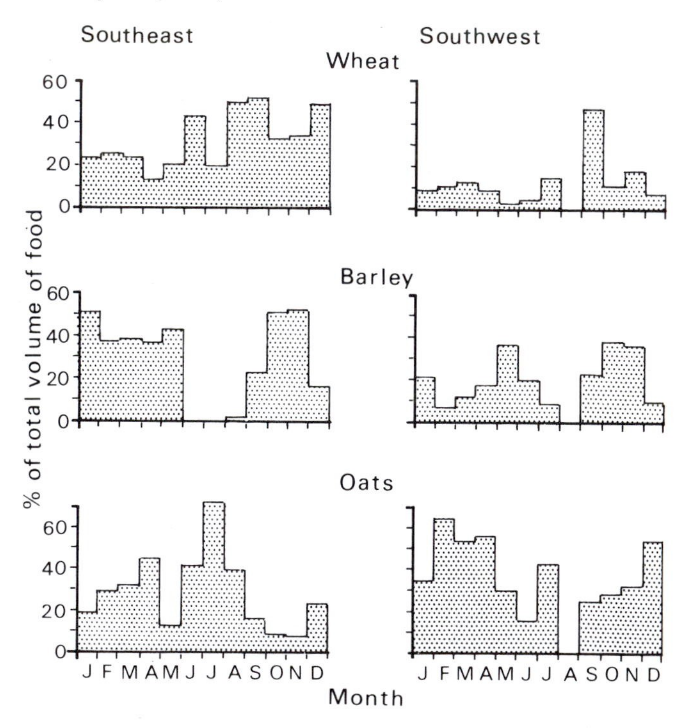

season, for birds may be found nesting at any time from the end of April to early October, but the period when most birds nest is relatively short. Murton (1958) found that the number of pairs breeding gradually increased up to July; breeding activity of the population was then rather constant until late September, when there was a sharp decline. The main breeding season corresponded to a peak in the availability of ripe cereals, the best food supply for nesting Woodpigeons. Breeding success similarly was dependent on cereals, with an early peak in success in March and April when breeding pairs could feed on the spring sowings. Success was very poor in May and June when no cereals are available and weeds, including weed seeds, are the principal food during these two months (Colquhoun, 1951). Green foods are especially taken at this time, particularly when the birds are feeding on buttercups (*Ranunculus* sp.). From July onwards cereals are again readily available and breeding success rises to a seasonal peak. Young Woodpigeons grow faster in August and September when food is plentiful and may leave the nest when only 16–20 days old, instead of the 33–34 days needed at other times of the year.

Since the early 1970s the steady spread of winter barley at the expense of spring barley has changed the pattern of availability of cereals,

Fig. 8.11. Woodpigeon breeding seasons recorded in England in 1957 by Murton (1958), and recorded in BTO nest record cards for Britain 1962–80.

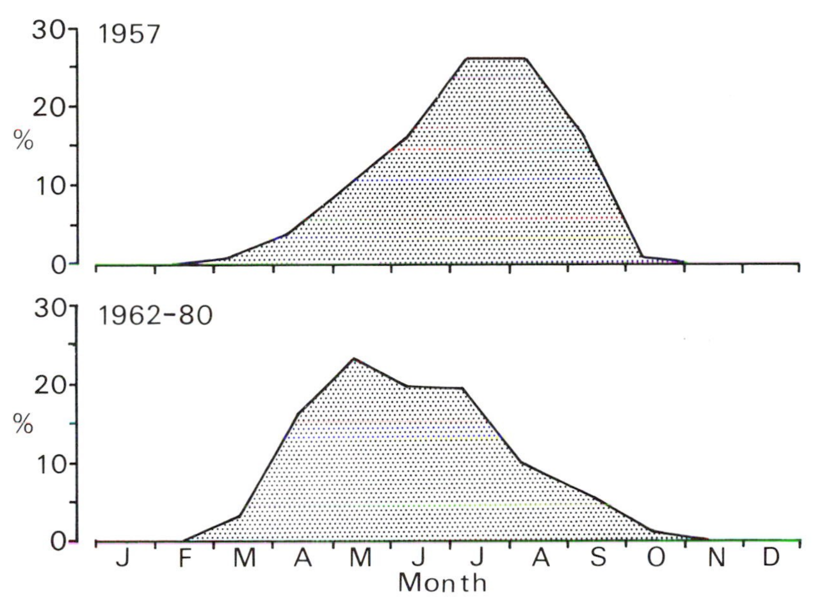

advancing the period of ripe grain by 3–4 weeks. Cereals have also become more widespread. Comparison of Murton's data for 1957 with our own for 1962–80 reveals three effects on Woodpigeon populations – on the timing of breeding, on its success, and on regional variations in success. Fig. 8.11 compares the breeding seasons recorded nationally at these two times and shows that laying has markedly advanced since Murton's work, as expected with earlier cereal ripening. BTO data are biased towards underrecording late season nests but the effect shown here is also evident within the BTO data alone, despite smaller samples; moreover, the advancement of laying dates has been fairly steady since the mid-1960s.

Breeding success has also changed with the expansion of cereals, from the 9–10 per cent of all nests fledging young in Murton's (1958) study and our estimate of nest success of 11.3 per cent for 1962–65 to success rates of 20–23 per cent between 1966 and 1980. Regional variations in breeding success also reflect the impact of cereal cultivation. Some 43 per cent of all winter barley is concentrated into eastern England, and here early broods (those fledging before 1 July) are both larger and more successful than in other regions, fledging an average of 0.33 chicks per nest against 0.07–0.21 in other regions. Late season broods in eastern England are not affected by the earlier availability of ripe barley and they are similar in success to late nests elsewhere (0.27 fledglings against 0.16–0.30). Taken together, these three lines of evidence indicate that changing cereal management has resulted in a significant alteration of the breeding biology of Woodpigeons.

Cereal habitats and Greenfinch populations

The expansion of cereal production has created new habitat which species such as Greenfinch have been quick to exploit and has also introduced new management regimes which have quite separately affected species formerly dependent on the weeds of arable lands, notably Linnet and, to a lesser extent, Stock Dove. Comparing the numbers of Greenfinches, Goldfinches and Linnets in various habitat groupings in the CBC shows that each species has moved in a different way – and in a consistent pattern – through tillage, pasture, sheep, cereals, and barley. In general, Greenfinch has shown a strong recovery since 1962–63 and then levelled off, whilst Linnet has mostly declined; the Goldfinch has been intermediate. The divergence between Greenfinch and Linnet is most marked in cereal-dominated areas. The pattern presumably reflects the differences in the feeding behaviour of each species. The Greenfinch makes far more use of cereal grains as food than either of the other species, so that the massive increase in cereal farming has actually provided it with an increased food resource. This view is supported by two other points. Examination of clutch sizes shows a significant difference in clutch sizes between areas dominated by cereal farming and those where roots and vegetable crops occupy an important proportion of the tilled area; clutch size is 14 per cent higher in cereal areas. Secondly, there is an increasing tendency for Greenfinches to dominate large winter finch flocks (p. 74).

Although Greenfinch numbers are favoured by the increase in cereal production nationally, their local distribution in winter (like that of other seed-eaters) may be determined by local management regimes. Comparison of winter flocks on two adjacent cereal farms in Sussex, one handled by ploughing and conventional cultivations, the other by direct drilling, showed that the direct drilling system was the more attractive during the winter months, holding 25 per cent more individuals per acre and twice as many species. The basic difference was that, despite the clean burn and chemical treatment, the direct drilling system still left a fair stock of waste grain and seeds which would have been lost by ploughing and which attracted and supported not only Greenfinches but also Yellowhammers, Corn Buntings, Reed Buntings, Chaffinches and both species of sparrows. Thus mixed flocks of *c.* 300 birds were present through most of November, December and January. So, even within cereal regimes, two different methods of growing the crops provide very different feeding resources and this must also be true of other forms of management. Sophisticated agricultural systems need careful management and are less tolerant of error than are traditional techniques. If their practitioners do not get this right, wildlife usually gains.

Herbicides and the decline of the Linnet

The increased use of herbicides which has accompanied modern cereal management lies behind the recent contrast in trends in Linnet and Goldfinch populations (Fig. 8.12). Both species have probably been influenced by changing levels of prosperity (which affects the standard of farm management) since the nineteenth century. Newton (1896) noted that both species had declined extensively and ascribed this decline as much to the reclamation of marginal land and improved agricultural methods reducing weeds as to the increase in bird-catching; some 2 938 000 acres (1 189 400 ha) were added to the area of crops and grass in Britain between 1867 and 1890. Parslow (1973) noted a general increase of Goldfinches in the twentieth century and a more limited increase in Linnets and several authors, e.g. Alexander and Lack (1944), noted that the Goldfinch had benefited particularly from the spread of thistles; this in fact resulted from the decline in agricultural management in the depression of the first 40 years of the century. That this decline in management standards also favoured Linnets is supported by examination of the records in Broad (1952). He analysed the incidence of weed seed in samples submitted for testing by the Official Seed Testing Station. His records for *Stellaria, Polygonum, Chenopodium, Brassica* (= *Sinapsis*) *arvensis, Rumex acetosa, Hypochaeris* and *Poa* – all listed as important food plants of the Linnet by Newton (1967) – show a uniform increase between the 1920s and 1950–51 in the number of crops tested in which they occurred – good evidence of an increase in distribution between the two periods. Increases in weed seed abundance within individual crops cannot be established in this way, however, since the searches for weed seeds were made on samples of cereal seed batches after cleaning, and methods of cleaning may have varied over time. Comparison of Collinge (1924–27) and Newton (1967) suggest that changes in farming practice were already influencing Linnets by the time of the Newton's study. Collinge noted that Linnets were widely regarded as pests of brassica seed crops, especially turnips, but Newton, writing 40 years later when the area of turnips grown had declined by 90 per cent, does not mention this food source at all. Collinge's records also give markedly higher value to the seeds of *Rumex* species than do Newton's. While docks and sorrels are not particularly easy weeds to eradicate, hormone weedkillers, available since the early 1950s, readily prevent them from seeding, and Sly (1977) reported that *c.* 9 million acres (3 640 000 ha) were being sprayed by these chemicals by 1969.

Over the last 25 years, however, Goldfinches and Linnets have differed in their responses to further agricultural changes. Fig. 8.12 shows the population trends for both species in a suite of counties dominated by cereal cropping. Until 1967 both species increased in density as they recovered from the 1962–63 winter but whilst the Goldfinch continued to increase over the following 8–10 years, Linnets declined and continued to decline up to 1983, by which time their numbers were lower than even in 1963. Goldfinches declined only recently. The differences in trends between the two species suggest ecological differences, and deteriorating conditions for the Linnet. Examination of the nest record cards for these same counties shows that Linnets have had greater difficulty than Goldfinches in rearing their chicks. Using the methods of Ricklefs (1969) to estimate the extent of chick mortality due to starvation, we estimated that the Linnet loses 36 per cent of its chicks in this way, whilst for Goldfinch the equivalent figure is only 27 per cent over a slightly longer nestling period.

Why have Linnets suffered such higher starvation mortality whilst Goldfinches have continued to prosper? Newton (1967) has shown that foods taken by Linnets are dominated by economically important weeds of arable farmland, particularly those of brassica crops, whilst Goldfinches are much less dependent on these foods. The Goldfinch shares only dandelion *Taraxacum* and the thistles with the Linnet and takes much groundsel *Senecio* and (in winter) burdock *Arctium*. Lack (1971) in fact regarded the two species as ecologically isolated in summer by diet. The

Fig. 8.12. Population trends in the Linnet (top) and Goldfinch (bottom) in cereal-dominated regions of England and Wales, 1962–83. The Linnet index was arbitrarily set to a value of 100 in 1962, but for Goldfinch small sample sizes (dotted lines) made 1967 a more convenient year to use as a baseline. Note the use of a logarithmic scale for population level.

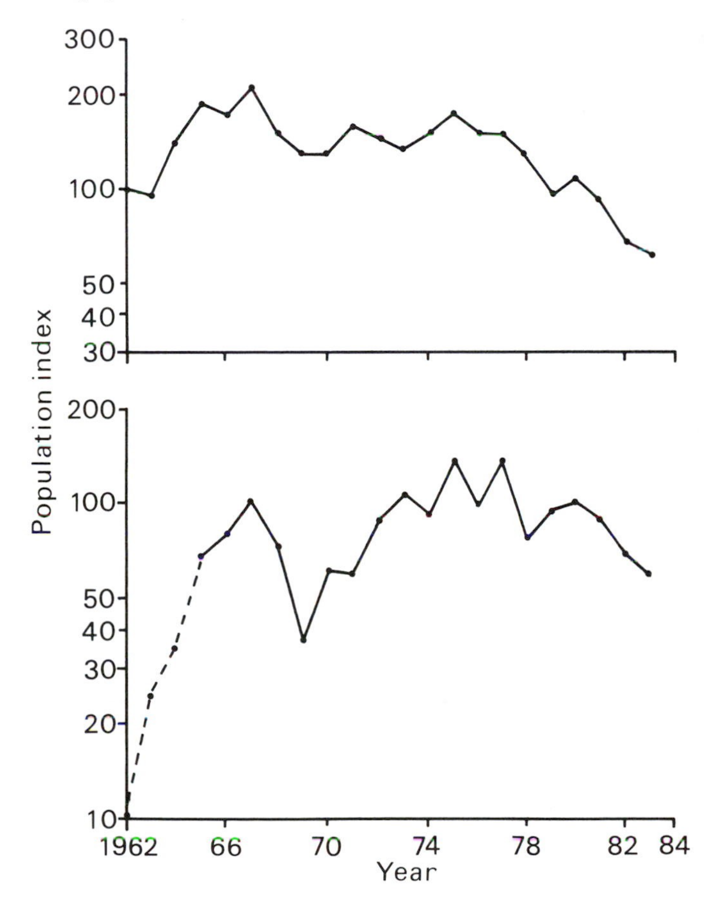

Linnet is thus much more vulnerable to the impact of herbicides which remove arable weeds from farmland.

Many of the food plants of Linnets are weeds of spring cultivations – spring cereals, clover and grass leys, and roots and vegetable crops (Broad, 1952; Gooch, 1963). Gooch's records show little change in the frequency of weed seeds in crop seed samples prior to 1960–61, except that *Rumex* species were less frequent and therefore probably declining in distribution. Even by 1970 such an easily controlled plant as charlock (*Brassica arvensis*) was still widespread on farms, present in 18 per cent of cereal seed samples taken during spring sowing (Tonkin & Phillipson, 1973). Other weeds were also frequent in farm-saved seed, e.g. *Polygonum aviculare* in 57 per cent of samples, *P. persicaria* 49 per cent, *Stellaria* 34 per cent, *Chenopodium* and *Poa* 17 per cent each, and *Rumex* 15 per cent. All were much scarcer in merchants' samples, reflecting both the better cleaning techniques and the stricter field hygiene of commercial seed crops but the farm-saved seed nevertheless constituted 28 per cent of all seed sown.

Fig. 8.13. Kale infested with fat-hen, a prime feeding site for winter finches.

Over the last 15 years this situation has changed drastically and many of the main food plants of the Linnet, particularly chickweed *Stellaria*, polygonums and, in brassica crops, fat hen *Chenopodium*, have come under chemical attack. Formerly these were rather difficult to control chemically or were in crops vulnerable to the chemicals then available (Fig. 8.13). Much attention has therefore been paid to them and a number of chemicals are now particularly recommended for these problems. Table 8.3 shows for each cycle of the Pesticide Usage Surveys the number of relevant chemicals listed in the ACAS Approved Lists and the areas sprayed in each crop. The intensity of modern weed control in field crops particularly favoured by Linnets is striking. There can be little doubt that this assault on its food plants is the main cause of the continuing decline of the Linnet over much of agricultural Britain. Analysis of the CBC records shows that, by 1980, the decline of the Linnet was general in all tilled areas but in

Table 8.3. *The availability of chemicals directed at the weeds forming the primary food of the Linnet*

	Number of chemicals recommended by ACAS	Crop	Area sprayed in various crops (000 ha)
1966–69[a]	19	Roots and vegetables	119
1972–74	27	Cereals	3149
		Roots and forage	190
		Vegetables	153
		Leys	129
		Total	3621
1977–79	31	Cereals	2927
		Roots and forage	222
		Vegetables	124
		Leys	513
		Permanent grass	490
		Total	4276
1981–82	32	Cereals	4489
		Roots and forage	374
		Vegetables	137
		Leys	487
		Permanent grass	1686
		Total	7173

[a] More specific information for 1966–69 is lacking but an estimated 3 643 724 ha were sprayed with hormone weedkillers.

primarily grassland areas its numbers were stable. The rapidly increasing use of herbicides in grassland shown by Table 8.3 must be expected to affect the populations in grassland areas in the foreseeable future.

The rapid spread of oilseed rape is the one feature of modern agriculture that may partly mitigate the effects of declining weed-seed stocks on Linnets. Like several seed-eaters, particularly Greenfinches, Linnets feed avidly on this crop as soon as the seeds are available.

In contrast to the Linnet, the Goldfinch has increased steadily over the period of the CBC, a continuation of the trend noted for most of the twentieth century. This partly reflects specific differences in food choices, for the Goldfinch takes more tree seeds and more invertebrates than the Linnet. Nevertheless, Newton noted that thistles (*Carduus* and *Cirsium*) formed a third of the annual diet and these are not only typical field and meadow weeds but are also a good indicator of the levels of farm management, having been controllable by hormone weedkillers since the early 1950s. Thistles, however, are particularly difficult weeds to eradicate completely because of the wide dispersal of their seeds, so that, unlike docks for example, a comparatively small area of unmanaged waste land will continually reinfest neighbouring farmland. Goldfinches also feed much more on the standing plant than do Linnets and are thus far better placed to exploit the scatter of thistles and other tall weeds still to be seen standing above many cereal crops. In addition, considerable food sources are available along the sides of many country lanes. These advantages have probably been sufficient to support the species' increase even today, partly because the area of marginal land within farmland is still, surprisingly, greater than in the late nineteenth century. Newton's account illustrates how well adapted Goldfinches are for exploiting even quite limited resources of this type.

The changing status of buntings

The pattern of increased herbicide usage may also have affected the status of Reed Bunting on agricultural land in Britain. In the 1960s this species spread from its previously typical wet marshy habitats into drier habitat more characteristic of the Yellowhammer (Kent, 1964). Weedy patches were particularly favoured during this expansion, and the bird concentrated along the weedier hedges and fields. Gordon (1972) showed that these drier habitats are usually taken up later in the season than are the preferred wet habitats and suggested that they were colonised only as an overflow habitat. On her Oxfordshire census plot, machine cutting often left the base of each hedgerow untouched, with the result that a rich ground-layer could develop and provide a source of seeds for late-nesting Reed Buntings. Evidence of yet further habitat overflow comes from two case studies of this species. Fig. 8.14 shows the distribution of Reed Bunting territories on Miswell Farm, a cereal farm in Hertfordshire. Here the early territories were established in scrub or in hedgerows but territories from mid-June

Fig. 8.14. Part of Miswell Farm, Tring, Hertfordshire showing a CBC plot of Reed Bunting territories in relation to hedgerow and barley crops. The stippled areas show the late (July) territories. After Williamson (1968).

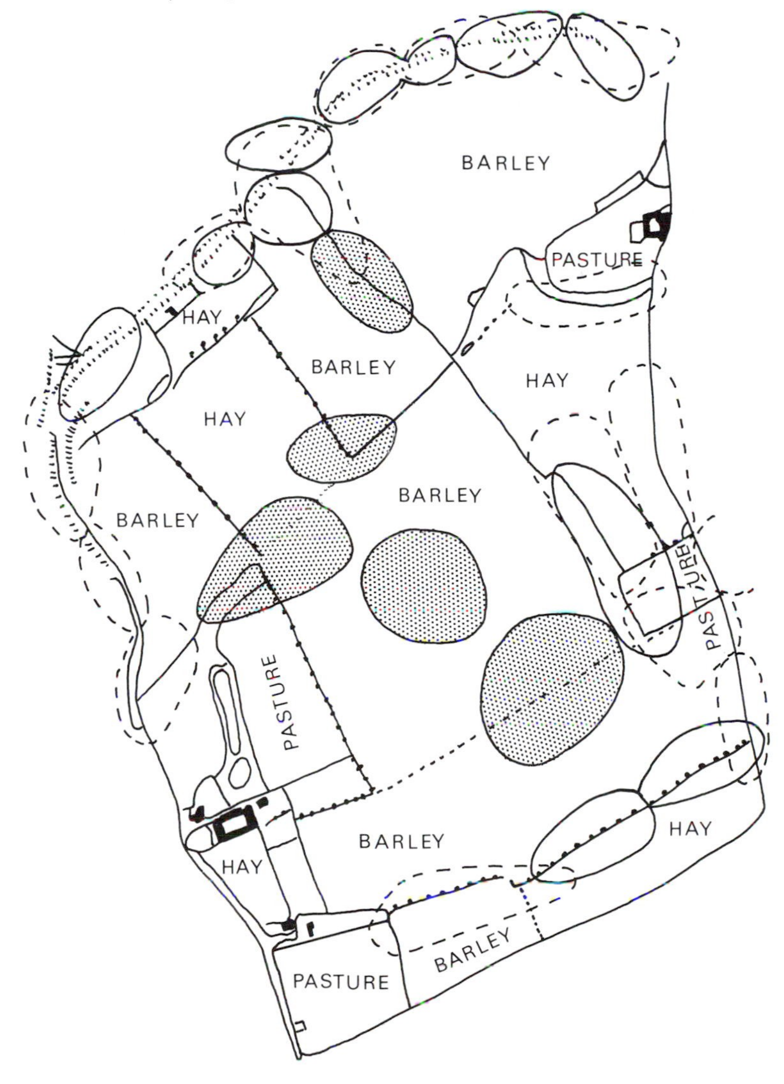

and July were established in fields of barley, a delay thought due to the birds having to wait for the barley (or its weeds) to grow to a suitable height (Williamson, 1968). Similarly, when fields of clover on Oakhurst were left weedy during the 1960s and early 1970s, Reed Buntings used them for

Fig. 8.15. Reed Bunting population trends in pastoral (top), cereal (middle), and tillage (bottom) areas of England and Wales. Dashed lines indicate small samples. Note that 1966 was used as the reference year (index = 100) except in cereal areas.

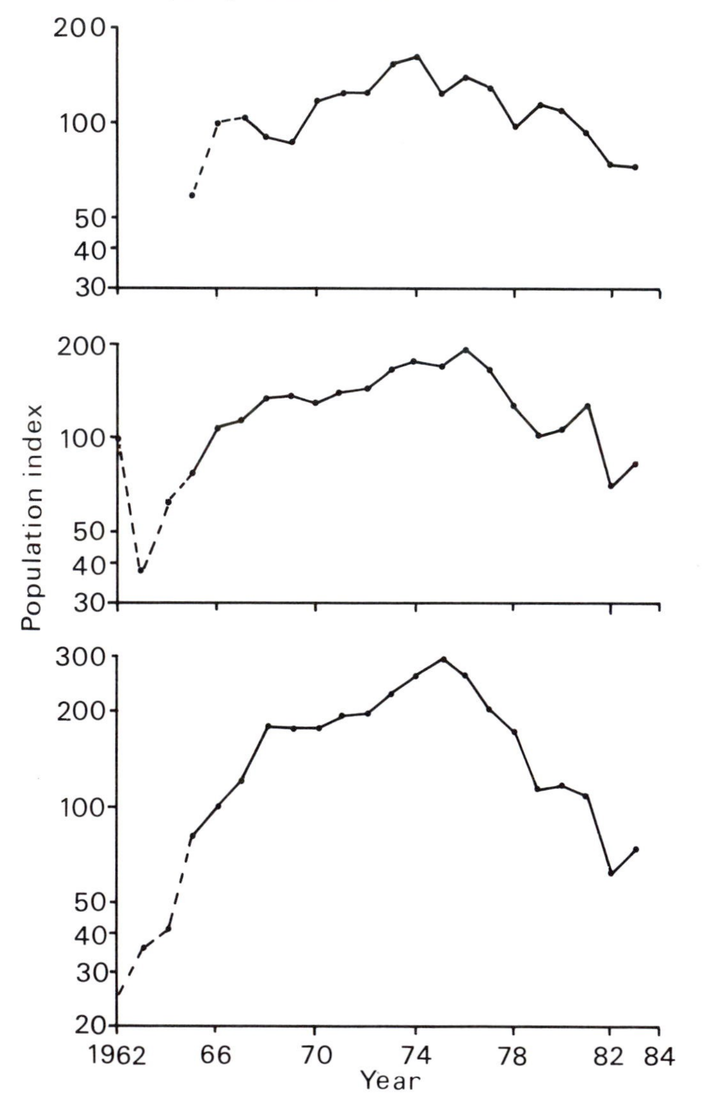

nesting. The national CBC shows steady population increase between 1963 and about 1974, an increase which presumably drove the expansion into the new habitats just described. Fig. 8.15 shows the CBC trends in pastoral, cereal, and tillage areas. All three have a similar pattern, of a recovery from the severe winter of 1962–63 until 1974 or 1975, since when there has been a steep decrease; population densities were also slightly elevated during 1966 and 1967. This marked similarity across the three areas requires a temporal explanation common to these subsets (and to others examined but not shown here). The steep increase in herbicide use that took place after 1974–77 constitutes the one change in agricultural practice that took place at the right time and scale to account for the trends in population shown. Oakhurst Farm again provides a case study, for Reed Buntings disappeared from the clover fields there once the spraying of undersown cereals with herbicides became the norm. Unlike the Linnet, however, the Reed Bunting is probably affected more by the change that herbicides produced on vegetation structure than by any effect on the availability of food. The destruction of weeds in clover swards, for example, removes much of the vegetation heterogeneity that provides the Reed Bunting with its nest sites.

Records for Reed Buntings nesting on agricultural land are rather few and distributed unevenly in time, so it is difficult to compare trends in breeding success within groupings of counties uniform in agricultural

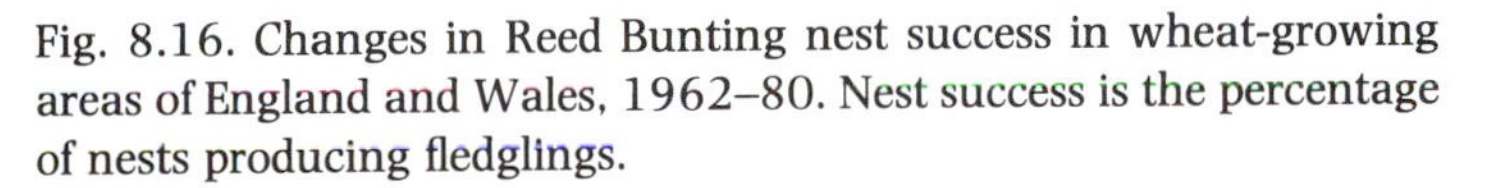
Fig. 8.16. Changes in Reed Bunting nest success in wheat-growing areas of England and Wales, 1962–80. Nest success is the percentage of nests producing fledglings.

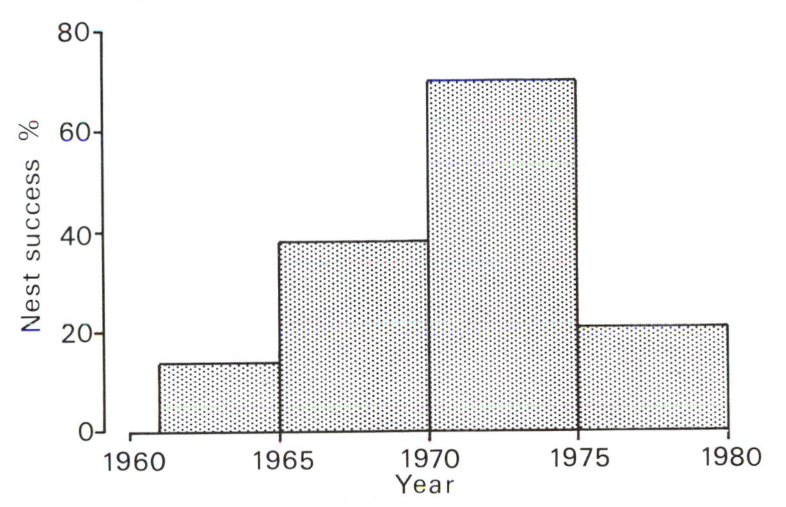

practice. Cereal and barley areas nevertheless show some small differences in overall success: in areas dominated by pasture average fledging success over 1962–80 was 30 per cent; in barley-dominated areas it was 34 per cent; and in the general cereal counties it was still higher, at 37 per cent. We therefore considered a larger, less restrictively defined, sample of data for wheat-growing areas and examined success through the CBC period. The results are shown in Fig. 8.16 and show that nest success increased as Reed Buntings spread into drier habitats but fell again in the latter half of the 1970s as herbicide use increased sharply.

The expansion of the Reed Bunting into drier habitats described above raises the possibility of increased competition with Yellowhammers. Lack (1971) considered that the two species were separated ecologically by habitat. Yellowhammers have declined in areas of tillage and pasture (and also in sheep-rearing areas) but they increased in cereal counties until the late 1970s, when they experienced a brief decline from which they are now emerging. The species therefore increased in the very habitat into which Reed Buntings were moving, so competition was probably relatively unimportant in shaping population trends.

Fig. 8.17. Yellowhammer population trends in areas of stock rearing in England and Wales: (top) where cattle are numerically dominant over sheep, and (bottom) where sheep are numerically dominant over cattle. Dashed line indicates small samples. Note the logarithmic scale and arbitrary choice of 1962 as the reference year (index = 100).

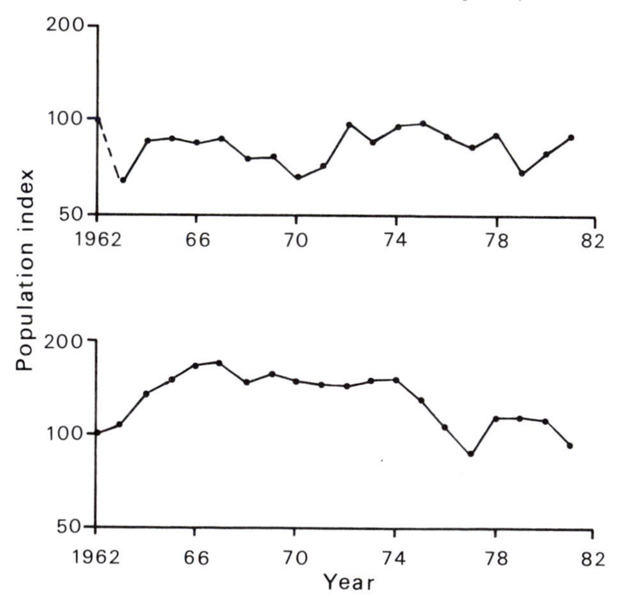

Yellowhammers make much greater use of grain than do Reed Buntings and also differ from Reed Buntings in being year-round residents. They thus depend to a greater extent than the latter on the pattern on agricultural practice in their local area. Yellowhammers do well in cereal areas in summer and early winter because they take more cereal grain but by January much of this food source (even under spring-sown cereal regimes) has disappeared and many birds find it difficult at this time to obtain adequate food to see them through the long cold nights (Evans, 1968). They are therefore dependent on whatever food concentrations may be available and at this time appear at garden feeders and other artificial sources of food (Glue, 1982). On agricultural land Yellowhammer densities are notably greater in stock-rearing counties (Chapter 4). In fact it is the management of stock that is most important here. Sheep may be kept and fed in folds in winter but more often receive their winter forage in the fields, often in the form of bales of feed dispersed across the fields from the back of a moving vehicle. Cattle, on the other hand, are normally fed in central stockyards in winter and this concentration of food sources has more effect on Yellowhammer populations than has sheep rearing. Fig. 8.17 shows how Yellowhammers have flourished in cattle-dominated areas but have declined in sheep-dominated areas.

Corn Buntings decreased in Britain with the decline in arable farming at the end of the nineteenth century and by 1930 had disappeared from

Fig. 8.18. Population levels of Corn Buntings in Britain 1962–83 in relation to barley acreages over the same period. Population level measured as the Common Birds Census index.

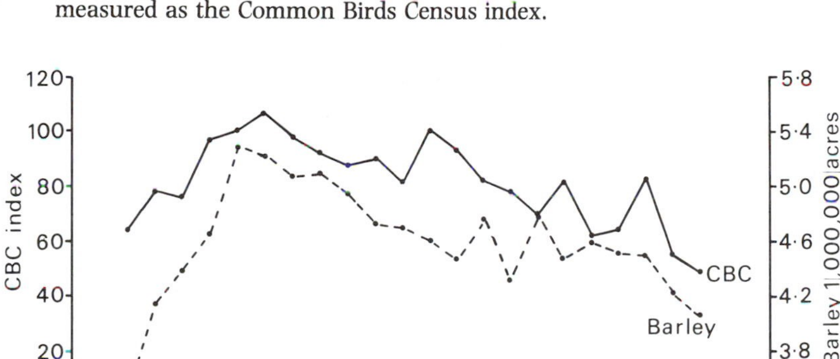

large areas of the country. A recovery started in the 1940s following the agricultural revival. Harrison *et al.* (1982), for example, document these trends in the West Midlands and show a continuing expansion into the 1960s, when numbers consolidated there. This increase is apparent nationally in the CBC results for 1962–68, after which densities have declined steadily (Fig. 8.18). The figure shows that these changes are correlated with changes in the acreage of barley in England and Wales. In the West Midlands Corn Buntings breed mainly in areas of arable or horticultural production, particularly on light, sandy soils, and the number of such breeding localities has increased since 1939 in step with increases in barley growing. This affinity between Corn Buntings and barley must be very close since it is apparent even within individual farms. In Fig. 6.10, for example, the size of the Corn Bunting population recorded varied in step with changes in the barley acreage there.

Population expansion in corvids

Considerable changes have taken place in upland agriculture. The increase in coniferous plantations and the increased stocking rates of sheep there have been particularly relevant to birds. The gross changes in bird communities that are brought about by afforestation were discussed in Chapter 5 but the increased density of sheep in the uplands may be more significant. Grouse and some waders, for example, have declined where the vegetation has been overgrazed (M. Pienkowski, personal communication). The increase in sheep numbers has been in response to the availability of headage payments which provide grant aid for each extra sheep kept on the farm. By this rather indiscriminate support farmers have been encouraged to increase their flocks at the expense of individual quality. The resulting increase in carrion should have favoured population expansion on the part of corvids, particularly the Carrion Crow and Magpie. Instead, Carrion Crow numbers have been rather stable in sheep-rearing areas, the surplus production apparently being exported to areas of cultivation (Fig. 8.19). Table 8.4 provides circumstantial evidence for this in the form of increased nesting success (particularly in egg success) in the sheep areas but constant success elsewhere, though the nest record samples for these subsets are rather low. The possibility of organochlorine sheep-dips contributing to the low egg success of the early 1960s must be kept in mind as an alternative to any increase in carrion; however, other areas were affected by organochlorines and show little evidence of corresponding improvements in breeding success for Carrion Crow. Either way, many more young have been fledged in the uplands than previously and in a strongly territorial species such as the Carrion Crow (Patterson, 1980) few of these extra young are likely to find vacant territories in their natal areas, thus creating a source for the colonisation of the lowlands. Indeed, Harrison *et al.* (1982) note an early increase in the West Midlands in the size of the flocks summering there. Such non-breeding flocks are often seen in species limited by territoriality.

Fig. 8.19. Carrion Crow population levels in areas of sheep (top), cereals (middle), and tillage (bottom) in England and Wales. Dashed lines indicate small samples. Note logarithmic scale and arbitrary choice of 1962 as the reference year (index = 100).

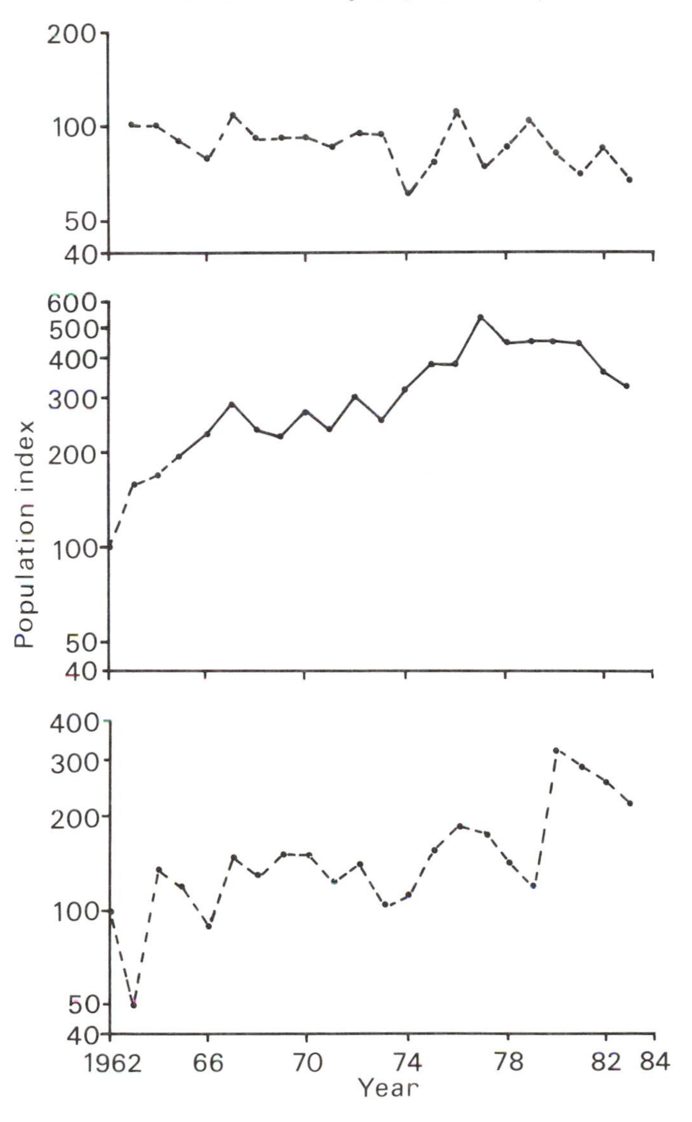

The Jackdaw has undergone significant changes in population status broadly similar to those of Carrion Crow and Magpie but they have been modified by habitat-related variations in breeding success and by changes in the availability of suitable nest holes. Nest success – the percentage of nests fledging at least one young – varies a little between areas dominated by pasture or by cereals or barley alone (Table 8.5), with pasture poorest and barley best. A big difference exists, however, between how well those eggs laid survive individually to produce fledglings: only 11 per cent of eggs laid in pasture areas yield fledglings, against 20 and 23 per cent in barley and cereal samples respectively. Much of the difference is due to the failure of individual eggs to hatch, suggesting that Jackdaws find it more difficult to produce good quality eggs when in pasture areas. Loss of nesting sites has been suggested to account for local decreases in Jackdaws, most recently in relation to Dutch elm disease (p. 144). In Essex, for example, a 1980 request for information on this species indicated decreases in many parts of the county, in several cases with a suggestion that loss of nest sites was responsible (Cox, 1984). However, Jackdaws feed extensively in summer on invertebrates in short grassland and Tapper (1981) has linked a *c.* 50 per cent decrease over the last 15 years in the numbers shot on East Anglian estates to a similar reduction in the amount of ley grassland present (Fig. 5.10). However, this explanation would not account for the poor breeding success in the pasture counties just described.

The increase in corvid populations has been aided in part by the reduced intensity of gamekeepering in the lowlands. This is perhaps best illustrated

Table 8.4. *Changes in breeding success*[a] *in Carrion Crows nesting in sheep-rearing and in cereal-growing areas between 1962 and 1980*

	Sheep-rearing areas			Cereal-growing areas		
	Eggs hatched %	Chicks fledged %	Nest success %	Eggs hatched %	Chicks fledged %	Nest success %
1962–65	7.7	61.1	4.7	(42.8)	—	42.8 max
1966–70	17.1	47.4	8.1	55.0	64.3	35.4
1970–75	49.3	45.9	22.6	43.7	46.9	20.5
1976–80	48.4	—[b]	14.0[c]	64.6	73.3	47.4

[a] Estimated using Mayfield's (1975) method for individual eggs and chicks.
[b] Chick mortality not known precisely enough to compute separate estimate.
[c] Estimated from total nest mortality, irrespective of when mortality took place, and so not fully comparable with the other time periods.

Table 8.5. *Components of breeding success*[a] *in Jackdaws nesting in different agricultural regimes*

		Barley areas	Cereal areas	Pasture areas
Incubation period	Nest survival[b] %	62.3	67.2	55.3
	Egg failure[c] %	3.7	7.8	30.7
Nestling period	Nest survival[b] %	73.5	61.5	71.0
	Partial brood loss[d] %	38.7	24.1	24.8
Nest cycle	Nest success[e] %	45.8	41.4	39.3
	Egg success[f] %	20.4	22.7	11.4
	Partial losses[g] %	25.4	18.7	27.9

[a] Estimated using the method of Mayfield (1961, 1975).
[b] Percentage of nests surviving the period.
[c] Partial losses only i.e. excluding losses of entire clutches to predation or desertion (see Ricklefs, 1969; O'Connor, 1984b).
[d] Excluding total brood losses to predation or desertion.
[e] Percentage of nests fledging at least one young.
[f] Percentage of eggs that resulted in fledglings.
[g] Percentage of eggs that failed in otherwise successful nests.

Fig. 8.20. Increase in Magpie population levels in cereal (top) and tillage (bottom) areas of England and Wales, 1962–83. Reference year (index = 100) set to 1966 in both cases. Dashed lines indicate small samples.

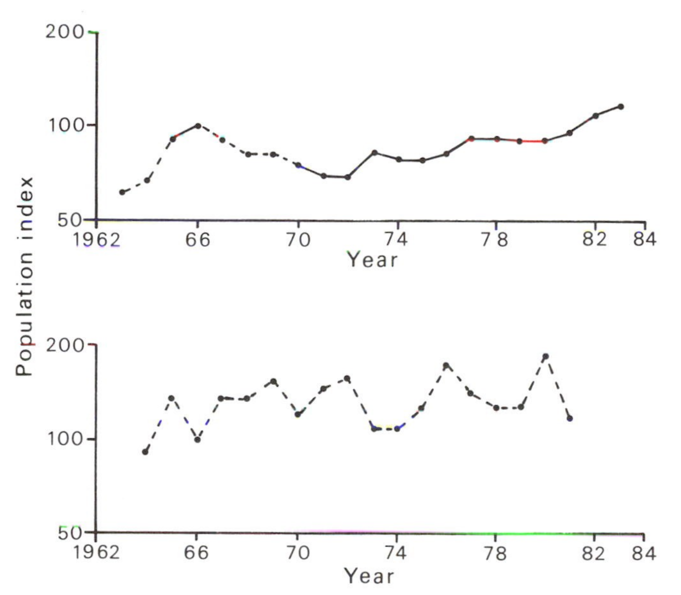

by the increase in the Magpie (Fig. 8.20), a species particularly persecuted by gamekeepers. Thus in Worcestershire Harthan (1946) noted that Magpie densities varied inversely with the number of gamekeepers in different parts of the county. Several other counties have recorded increases attributed to reduced persecution in recent decades, e.g. in Surrey (Parr 1972), in Leicestershire (Hickling, 1978), and in Derbyshire (Frost, 1978). The same is true of the Carrion Crow. In Derbyshire, for example, the Carrion Crow has historically (and currently) been most numerous on the high ground where game preservation has not been practised, and numbers also increased during the two world wars when gamekeepering was reduced (Frost, 1978). Nevertheless, without the high source populations in the uplands the recolonisation of the lowlands on the cessation of persecution could not have proceeded as rapidly.

Overview

Of the 13 species discussed in detailed case studies in this chapter, five have probably been influenced by crop management, four by changes in the timing of tillage, four either by changes in the area of grass or tillage (which are often linked) or cereals, and two each by stock management, grassland management or herbicides. Clearly bird populations have been as strongly influenced by such management changes as by the habitat changes that have dominated our thinking in this area in the past. Management has thus significantly changed the farmland environment experienced by birds in the past 40 years, principally by increased technical efficiency. It seems worth making the general point that in every type of farming system, no matter how traditional, wildlife thrives on inefficiency. Overgrown hedges, weedy fields, undrained pastures – all of which improve the feeding resources available to birds – are simply symptoms of inefficient management. Such inefficiency is not necessarily a reflection on the competence of the farmer; it is as likely to reflect his assessment of the economics involved. But, as a broad generalisation, each of the impact points discussed above reflected the application over large regions of the countryside of methods which give farming a greater technical efficiency. This point provides a strong conservation argument in favour of small farms. Avian conservation benefits most from having larger numbers of smaller farms than vice versa, for individual variation in levels of ability and interests are a prime source of diversity in countryside. Mixed farms are always better habitats for birds than specialised ones. Nowhere is this more striking than in Lapwing populations which find their largest stronghold on grassland managed within arable rather than within pastoral areas.

9: Pesticides and pollution

Chemical farming is the one completely new element which modern farming has introduced into the countryside. Pesticides (the generic term properly used to cover the whole range of farm herbicides, fungicides and insecticides) have three series of effects. They kill plants and animals, often selectively, thus directly influencing the environment and its ecology; selectivity may be as significant ecologically as elimination. Secondly, they may change food and habitat resources, e.g. by removing weed species on which insect populations depend, and thus indirectly affect the numbers or distribution of forms immune to direct effects. Thirdly, they also allow the farmer much greater freedom to select his crops and rotations. They thus bear strongly on the basic form of the farmland habitat and have been a crucial factor in the modern pattern of intensive units specialising in certain crops (discussed in detail in Chapter 5). Cereal farming provides a particularly clear illustration of this. There are only two ways of controlling weeds in cereal farming, by cultivations which kill weeds mechanically or by herbicides which kill weeds chemically. Intensive cereal farming, especially that based on autumn-sown crops, simply does not provide time to control weeds by cultivation; this applies particularly to grass weeds, the most important check in such farming. For this reason the great expansion of the spring barley area from 1950 to about 1967 (which may be taken as an index of the expansion of continuous cereal cropping) was made possible by the rapid development of selective herbicides which gave good control of a wide range of troublesome cereal field weeds, though not of grasses. These latter were still controlled by the longer period of cultivation provided by spring sowing. The major switch to winter cereals, which started very rapidly in the early 1970s, has been similarly associated

Fig. 9.1. Number of sprays available on the ACAS list 1951–85, for the control in cereals of: weeds (A–G), insects (H–J), molluscs (K), and fungi (L–N). The specific groups are: A, general weed control; B, general weed control in undersown crops; C, control of perennial grasses in stubbles and crops; D, 'difficult' weeds and cleavers; E, annual grasses and wild oats; F, pre-emergent herbicides; G, growth regulators; H, insect pests; I, summer aphid attacks; J, autumn aphids; K, molluscs; L, general fungicides; M, rusts; N, eyespot, For 'difficult' weeds see p. 175.

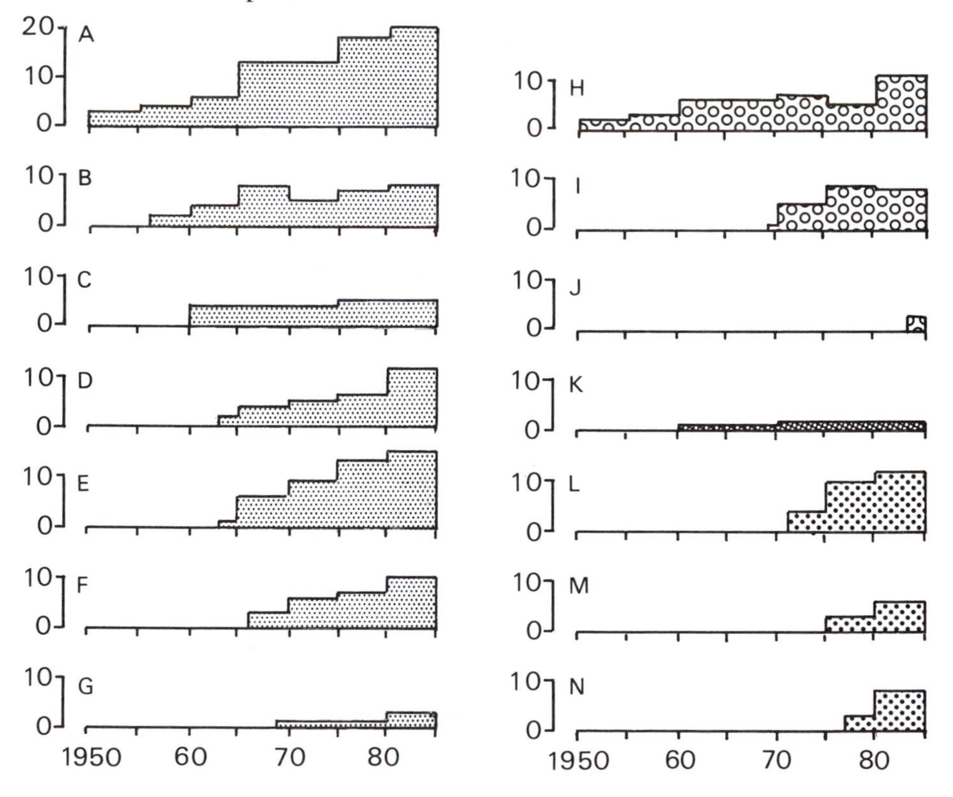

with the development of herbicides capable of selectively controlling grass weeds and wild oats, often in pre-emergent applications.

Fig. 9.1 shows the increase since 1951 in the number of crop chemicals on the Approved List of the Agricultural Chemicals Approval Scheme (ACAS) that were available for use against various pest categories in cereals. For herbicides the figure shows, first, a steady increase in the variety of weeds attacked, and second, an increase in the variety of chemicals available within any category. Hence although early success came in general weed control, some species such as polygonums and cleavers were more difficult to control; these were in turn made the target of further chemical development. Both the variety and abundance of weeds have therefore fallen, with consequences for seed-eaters like Linnet (p. 171) and Stock Dove (p. 211). Insecticides have increased in variety since the 1950s, particularly aphidicides since 1970. Fungicides have come into major use only in the latter part of the 1970s. Fig. 9.2 shows the seasonality of use in cereals of these pesticides in 1981–85. Spring and late autumn have always been the times of spraying but the variety and volume have increased markedly and spraying has spread to all months, although not necessarily in a single system. Hence both weeds and insects can come under chemical attack throughout the year.

Table 9.1 extends the picture shown in Fig. 9.1 to cover the general growth in chemical farming from 1955 to 1985. It is based on the ACAS Approved List of MAFF and is not, therefore, necessarily exhaustive. But it is a consistent record covering the whole period of agrochemical use and must give a clear and accurate reflection of the growth of chemical usage. The table is based on chemical formulations rather than on manufacturers' products, since the latter additionally reflect the extent of competition for the market. The summary of usage in Table 9.2 is for four principal crop

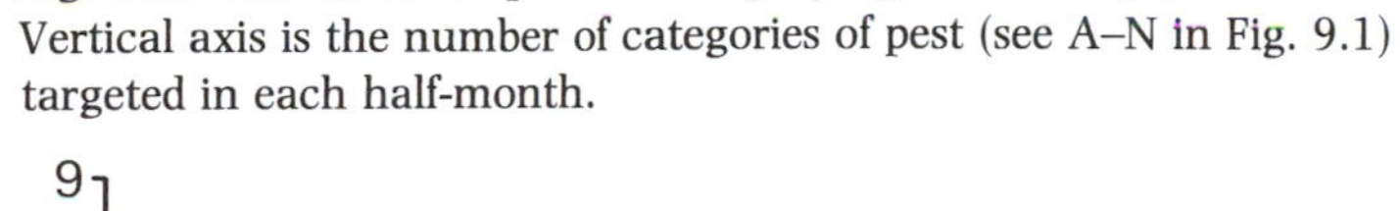
Fig. 9.2. The seasonal pattern of spraying cereal crops, 1981–85. Vertical axis is the number of categories of pest (see A–N in Fig. 9.1) targeted in each half-month.

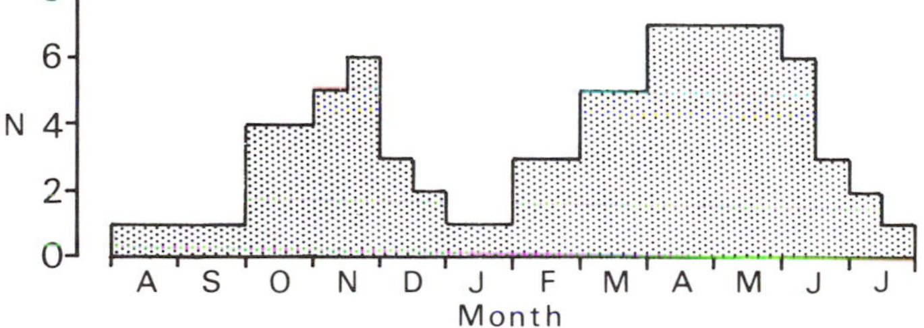

Table 9.1. *The growth in types of chemical used on farms in Britain, 1955–85*

	ACAS list	Insecticides[a]			Herbicides			Fungicides		
					N of chemicals used in			N of chemicals used in		
Year	N of chemicals	N of chemicals OC[b]	OP[c]	Carbamates	Total	Cereals	Roots[d]	Total	Cereals	Roots[e]
1955	37	3	3	2	5	5	1	4	1	4
1960	53	6	5	3	13	11	1	5	1	4
1965	101	11	19	5	31	16	9	20	1	9
1970	136	13	24	7	52	24	20	26	1	13
1975	178	10	36	14	69	43	32	41	8	13
1980	198	11	30	19	79	52	36	48	14	18
1985[f]	199	6	27	22	80	33	39	52	18	24

[a] Including seed dressings, acaricides, and nematocides.
[b] Organochlorine compounds.
[c] Organophorus compounds.
[d] Roots taken to include potatoes, sugarbeet, brassicas for vegetables, fodder roots and brassicas and oilseed rape. Many herbicide applications for these crops are pre-emergent or pre-planting.
[e] Primarily against blight in potatoes.
[f] The Crop Chemicals Guide for August 1984 which includes chemicals without ACAS approval lists 40 herbicidal chemicals for cereals and 43 for roots and lists 21 fungicidal chemicals for cereals and 30 for roots.

groups and is based on the Pesticide Usage Surveys conducted by MAFF since 1966. The tables are a striking demonstration of the extent to which chemicals now dominate the raising of crops. Each of the main chemical groups listed is examined in more detail below.

These summaries of pesticide usage have considered only tillage crops. But insecticides and more particularly herbicides are widely used in ley grass. Here, however, their impact, if any, is virtually impossible to separate from the major habitat change involved in reseeding, particularly where old pastures are replaced by high output rye-grass leys. As already noted, a continuous pattern in the development of chemical farming has been the change in the methods of application. Research in this area continues and is now largely aimed at securing placement of the chemical exclusively on the crop or target plant, reducing drift and reducing dose rates. While the successful application of such research will reduce the quantities of pesticides used, it should be remembered that the efficiency of the chemical and the area and scale of use are the relevant ecological factors. The reduction or elimination of drift into non-target areas would undoubtedly be a major advance. Newton & Haas (1984) stress the importance of drift in the wide dispersal of organochlorines throughout the environment. Drift has probably also an inoculant effect, which exacerbates the problems of resistance in target organisms. Yet the indirect effects of chemicals may be as serious to farmland ecology (Bunyan & Stanley, 1983; Hardy & Stanley, 1984). Some carbamate compounds are highly toxic to earthworms (Edwards, 1984), for example, and their deployment on farmland could therefore deprive soil-feeding birds such as Rooks of a most important food

Table 9.2. *Summary of total usage of farm chemicals in England and Wales, 1966–82*

Pesticide survey cycle[a]	Total area sprayed 000 ha	Total used tons AI[b]
1966–70	10667	11062
1971–74	12783	22061
1975–79	15795	25497
1981–82[c]	20841	19936

[a] No relevant survey took place in 1980.
[b] Active ingredient.
[c] Survey cycle incomplete, reports published covering only cereals, vegetables, sugarbeet, potatoes, fodder roots and oilseed rape.

source. Little evidence is available on such effects in bird populations. Although we can show population decreases by several species on CBC plots in areas of heavy carbamate use, the areas concerned have experienced several other agricultural changes which might account for our results. The question is one which probably requires specific investigation rather than reliance on the surveys analysed here. Another aspect of the indirect effects of chemicals is how the palatability of crops may be affected by spraying or by take-up of soil compounds. Birds may be averse to eating plants treated with carbamates (Green, 1980), though here the greater problem seems to us to lie in the great toxicity to vertebrates of these chemicals (Hardy & Stanley, 1984). Stanley & Hardy (1984) and Potts & Vickerman (1974) provide recent reviews of the indirect effects of agricultural chemicals.

Fungicides

Fungicidal seed-dressings in some form have been used for a very long time by farmers; in the early nineteenth century Col. Peter Hawker was complaining of the effects on partridges of blue vitriol applied to seed corn. Modern fungicides are used at two stages of crop development. Seed dressings are used at planting, to control seed- and soil-borne diseases, and foliar fungicides are used in the later stages of growth, to control mildews, rusts and other diseases of the growing plant. The availability of these chemicals has encouraged cropping changes – for example, in cereal farming a switch to higher value wheat crops, which are more susceptible than barley to the disease problems of continuous cropping. As a result the area of wheat grown in Britain increased between 1974 and 1980 nearly four times faster than that of cereals. The Pesticide Usage Surveys confirm that intensive wheat growing involves a higher use of pesticides than does the farming of other cereals. The pattern continues to change as winter barley also increases, winter barley systems being very similar to those for winter wheat.

The most widely used substance since the 1950s, however, has been organic mercury, which is used on most cereal crops and on many root and vegetable crops. The total area of use in England and Wales was 2.42 million ha during the mid-1970s or 24 per cent of total farmland area, but use in the 1980s seems to have declined sharply. Although quite high concentrations of mercury have been found in some bird species, notably Herons, Kestrels, and Barn Owls, the extent to which such concentrations are associated with increased mortality is unknown (Cooke, Bell & Haas, 1982).

The use of foliar fungicides has been the major growth area in agrochemicals in the 1980s, particularly the enormous increase in their use in cereals (Table 9.3). This has led to an increase in the spray area of fungicides of *c.* 265 per cent since 1977; in some cereal-growing counties this represents growth from *c.* 9 per cent of land area to *c.* 30–50 per cent. The implications of this change are discussed under the section on insecticides, for research carried out by the Game Conservancy has shown that fungicides can have important indirect effects, and some more limited direct effects, on cereal insect populations (Potts, 1984a).

General trends in insecticide use

Insecticides may be used both as seed dressings and as sprays at any stage of crop growth; they are also used in stock farming against such pests as

Table 9.3. *The use of fungicides on farms in England and Wales in selected crops*[a]

Pesticide survey cycle	Crop[b]	Area sprayed[c] ha	Area sprayed as % of area grown[c]	Active ingredient tons
1966–70	Orchards	594585	1049	1089
	Vegetables	88915	41	*c.* 40
	Roots[d]	188259	99	765
	Cereals	NIL	NIL	NIL
1971–74	Orchards	487389	968	699
	Vegetables	99545	52	*c.* 40
	Roots	597087	121	695
	Cereals	571387	18	394
1975–79	Orchards	441631	1029	565
	Vegetables	41089	16	36
	Roots	616908	116	833
	Cereals	976825	31	568
1981–82[e]	Vegetables	53575	27	47
	Roots	677160	115	528
	Cereals	5246662	152	2399

[a] Seed dressings are excluded (see text).
[b] Brassicas for human consumption are here included under vegetables.
[c] These figures take account of repeated spraying of some or all of each crop within the survey year for that crop. See Appendix 3 for details.
[d] Roots include sugarbeet and potatoes only in the first cycle, fodder and forage being included with cereals in the survey; otherwise sugarbeet, potatoes, fodder roots and brassicas and oilseed rape.
[e] The cycle for the 1980s is incomplete (see note to Table 9.2).

warble fly in cattle or blowfly in sheep. In general, however, there are well-defined major pests and risks in each principal crop. In cereals, for example, these are likely to be wheat-bulb fly, leatherjackets or aphids. The last may damage the crop directly or may do so indirectly by introducing disease. The importance of each varies regionally and may also vary with cultivations and rotations. Wheat-bulb fly, for example, occurs only in parts of eastern England and Scotland. Among brassica crops, those which are drilled to a stand are the most vulnerable to economically damaging attacks by soil pests, because of the low seed rates. Leatherjackets are most likely to cause problems in cereals where these alternate with grass.

One result of these general patterns is that comparatively few chemicals are in really widespread use. For example, DDT and HCH formed 88 per cent of the total tonnage of organochlorines used during 1975–79 on the crops listed in Table 9.5 below. Similarly, nine out of about 36 organophosphorus compounds comprised 75 per cent of tonnage used and just four of them accounted for 57 per cent of use. Again, four out of ten carbamates totalled 91 per cent of the total tonnage.

Table 9.4 summarises the number of products derived from these basic compounds and approved for use on crops under the ACAS scheme; it also lists methods of application. Table 9.5 lists tonnage and area of use for selected crops where recorded since 1966. Several points emerge. First, the overall number of products reached a peak about 1970, increasing by 32

Table 9.4. *The use of insecticides on farms in Britain, 1955–85, showing number of products and methods of application*

	Number of products[a]				Methods of application			
Year	OC[b]	OP[c]	Carbamates	Other	Dusts	Granules	Liquid sprays[d]	Baits
1955	56	14	nil	64	53	nil	65	nil
1960	73	15	nil	41	36	nil	78	6
1965	97	36	2	30	19	2	126	6
1970	96	51	7	26	9	15	120	14
1975	43	78	14	20	3	25	113	13
1980	32	72	16	20	1	28	94	14
1985	18	53	18	24	1	18	59	9

[a] Seed treatments are included but not tar-oil washes nor treatments for stored produce.
[b] Organochlorine compounds.
[c] Organophosphate compounds.
[d] Liquid sprays includes wettable powders applied as sprays.

per cent between 1955 and 1970 and then declining by 42 per cent up to 1985. This may reflect partly the introduction and use of the more effective carbamates and partly some rationalisation within the agro-chemicals industry. Secondly, Table 9.4 illustrates the switch away from organochlorine compounds. These declined from 42 per cent of approved products in 1955 to only 16 per cent in 1985. At the same time the area over which they are used has fallen steadily, particularly since 1977–79. An important part of this change has stemmed from the reductions in orchard area and in the use of DDT in orchards. In the 1970s, however, DDT use on field vegetables increased substantially, partly offsetting this gain. Today, DDT is banned and HCH remains the only organochlorine in general use, mainly as a seed dressing in cereals. Thirdly, the table shows important changes in the technology of application. In 1955 45 per cent of approved insecticides were applied as dusts; by 1985 only one dust was approved but 17 per cent were applied as granules incorporated into the soil. Some are systemic in action and absorbed through the plant's root system – and insects such as bees, for example, are less at risk to them than to dusts or sprays. But the carbamates are largely granular pesticides used against soil pests and are much more toxic to soil invertebrates (particularly earthworms) than are either organochlorines or organophosphates (Edwards, 1984).

The Pesticide Usage Surveys are based on one year's records for each crop in each cycle and the totals may therefore be influenced by annual variations in pest populations. Nevertheless, Table 9.5 shows a consistent increase in chemical use. Some of this increase stems from changes in cropping and techniques, coupled with the steady decline in labour employed. In addition to the switch to winter wheat already noted, modern rotations of cash crops, the modern fashion for early drilled cereals, precision drilling of root crops, and the decline in cultivations all appear to increase the use of insecticides; only precision drilling certainly does not, because of the sharp decline in seed rates, and therefore seed dressings, needed.

As with fungicides, it is in cereal farming that the greatest increase in insecticide use has occurred during the past decade, mainly since 1977. Prior to this cereal crops received only limited insecticidal spraying, and comparatively high levels of insecticide exposure were confined to areas or counties in which there were substantial areas of root crops, vegetables or orchards. By 1984, however, the use of cereal aphidicides alone reached 47 per cent (1.85 million ha) of the acreage grown, against 17 per cent in 1977 (see Fig. 9.3). This change has led to a major extension

Table 9.5. *Summary of use of insecticides in selected crops in England and Wales*

Pesticide survey	Crop[a]	Area grown 000 ha	Amount sprayed[a] 000 ha	Amount used[b] tons AI[c]	% sprayed[a] area by OC[d]	OP[e]	Carbs[f]	% tons used by OC[d]	OP[e]	Carbs[f]
1966–70	Cereals	3335	1827	181	95.9	4.1	nil	86.4	13.5	nil
	Roots	356	401	135	47.2	52.3	0.5	12.5	87.2	0.3
	Orchards	57	180	125	53.4	42.5	4.1	69.3	24.2	6.5
	Vegetables	214	170	92	41.7	58.3	nil	44.1	55.9	nil
1971–74	Cereals	3184	853	157	88.9	10.6	0.5	76.1	23.5	0.4
	Roots	518	516	154	50.6	48.6	1.8	32.9	66.2	0.8
	Orchards	50	173	113	31.7	61.1	7.2	42.8	40.4	16.8
	Vegetables	192	245	151	26.9	73.1	nil	20.5	79.5	nil
1975–79	Cereals	3138	1607	333	63.3	18.6	18.1	36.0	40.3	23.7
	Roots	530	578	335	23.6	46.8	29.5	20.9	33.2	45.8
	Orchards	43	154	86	22.2	64.2	13.5	27.3	46.4	26.3
	Vegetables	251	403	293	19.4	73.0	7.6	14.6	72.9	12.5
1981–82	Cereals	3444	2164	357	46.8	24.7	25.3	18.9	44.9	32.1
	Roots	587	722	388	35.1	23.3	53.6	16.0	20.0	63.9
	Vegetables	194	364	306	12.7	69.6	12.8	9.7	78.9	11.0

[a] Definitions as in Table 9.3.
[b] Seed dressings are often dual purpose and weights have been adjusted wherever possible. Unspecified treatments are omitted.
[c] Active ingredient.
[d] Organochlorine compounds.
[e] Organophosphate compounds.
[f] Carbamates.

of the area of the countryside exposed to insecticidal sprays, particularly in counties where tillage has always been dominated by cereals. This extension may amount to about 20 per cent of the land area of such counties.

The Game Conservancy's work since 1969 has established major changes in the insect populations of cereal fields. Three groups of insects have experienced steep declines in abundance: (1) species that feed upon field weeds; (2) the polyphagous predators, i.e. species that feed on a wide variety of other insects; and (3) species that feed on fungi. These last – a group which includes some major aphid predators – have shown most decline (e.g. five species of *Tachyporus* by over 95 per cent in Sussex (Potts,

Fig. 9.3. Recent trends in the use of aphidicides on cereals in England and Wales. Data for 1974, 1977 and 1982 from MAFF surveys. Data for 1984 from the British Agrochemicals Association Handbook 1983–84.

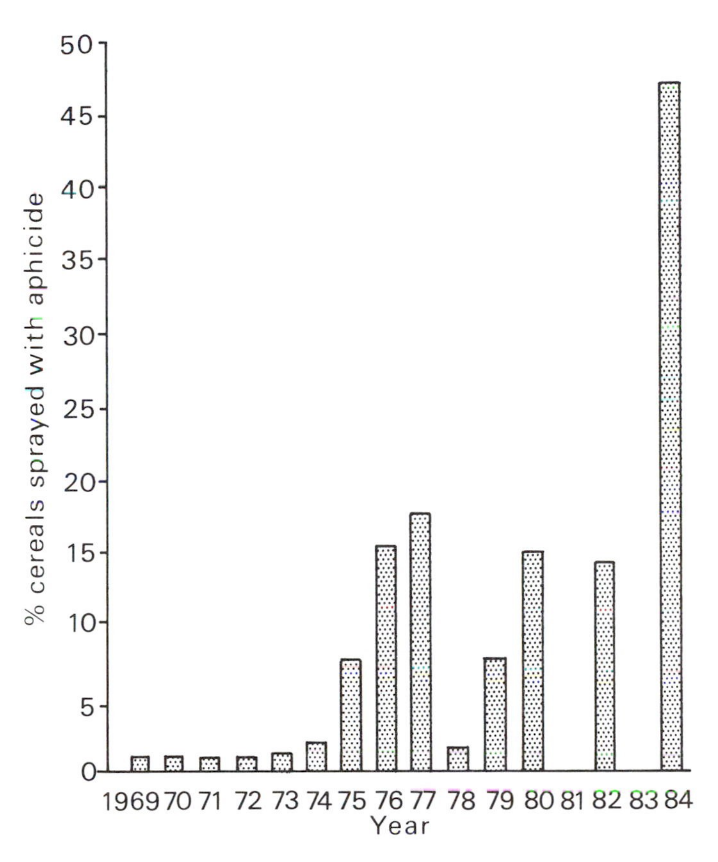

1984a)) and the least ability to recover. In contrast, species that feed on cereals have remained relatively constant. It seems likely that the wide-scale use of foliar fungicides and herbicides in cereals (see tables) has made a major contribution to the emergence of aphids as a serious cereal pest in the past 10 years. It is a classic case of pesticide use creating further use and it is probably not a coincidence that the real problems with aphids have appeared with the widespread application of foliar fungicides. This coincidence shows clearly in a comparison of Tables 9.5 and 9.3.

The Grey Partridge is one species for which the effect of insecticides and the insecticidal effect of herbicides have been particularly well studied (Potts, 1980, 1981a). In this species chick survival is a principal factor determining population changes (Southwood & Cross, 1969). Potts (1980) showed that chick survival largely depends on the number of arthropods available in cereal crops in June. He and his colleagues found that chick survival increased by 1.5 per cent for each extra plant bug (Heteroptera) per m^2 of cereal crops; by 4 per cent for each extra sawfly larva (Hymenoptera: Symphyta) per m^2; and by 3 per cent for each leaf beetle

Fig. 9.4. The effect of spraying of cereals with insecticides on the survival of partridge chicks. Most spraying was in the 0–20 index levels and within this area means and 95 per cent confidence limits are plotted. The five highest levels are individual estates. After Potts (1981a).

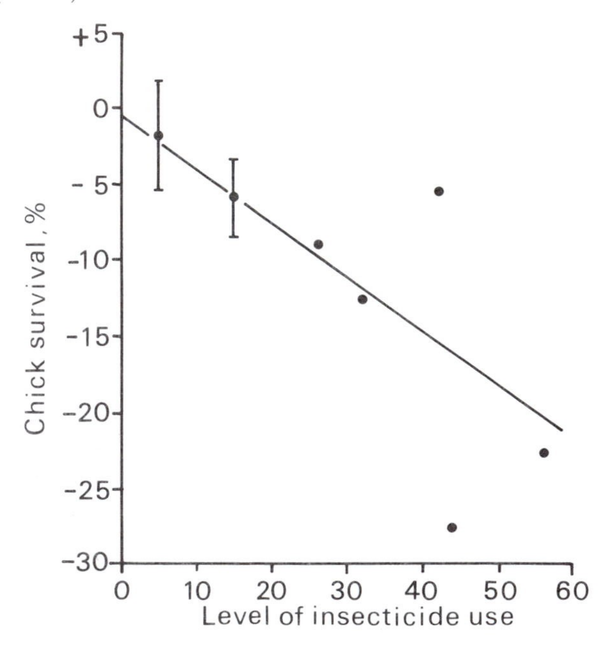

(Chrysomelidae) per m². Potts' conclusions were based on analysis of chick survival over a number of years, so there is some residual possibility that insect numbers may not be the critical factor *per se* but may be merely a symptom of the influence of some other, unmeasured factor such as weather or climate. However, Green (1984), working over a number of sites within a single year, also found that insect numbers were important for chick survival. These insect populations are influenced by several features of modern cereal management other than chemical use, for example, the decline in undersowing and the practice of straw burning, but Potts (1984a) nevertheless was able to show that herbicides were responsible for a 50 per cent reduction in the numbers of the most preferred insects. The direct effects of insecticides are, in contrast, less important but the use of aphidicides in winter wheat caused a reduction of up to 20 per cent in chick survival (Potts, 1981a). Although the use of aphidicides tends to fluctuate (Fig. 9.3), the general trend is for an increase over time and any such increase markedly reduces chick survival (Fig. 9.4).

That pesticide use is the main factor influencing chick survival is indicated by another feature of chick behaviour, a preference for feeding at the edge of fields (Green, 1984). Chick survival in fact improved in fields where the headland was experimentally left unsprayed to improve food stocks there (Rands, 1985). No matter what standard or form of management is practised in a cereal field, the edge and headland are always most likely to provide the best feeding for chicks. The young birds can work the interface between hedge and field and the proximity of the edge enables plants and their associated insects to move into the field at the headland.

Consequences of organochlorine use

Fig. 9.5 illustrates the use of organochlorines on farms in England and Wales in the early 1960s on a regional basis. In Fig. 9.5(*a*) the use of dieldrin sheep-dips in 1961–63 is shown, assuming that a total of 500 tons was used in Britain; Fig. 9.5(*b*) shows the use of dieldrin, aldrin and heptachlor (the cyclodiene group) in arable crops in 1962–63; and Fig. 9.5(*c*) shows the total use of organochlorines in arable crops and orchards for the same period. These maps are based on calculations made from MAFF (1964), which show a fairly wide range in the amounts of active ingredient used in the various treatments. Thus the tonnages mapped should be regarded as fairly rough estimates. Even so, organochlorine usage on farms at that time clearly was extensive and very uniform in England and Wales.

It is difficult to assess the effect of sheep-dips. In upland areas the

Fig. 9.5. The use of organochlorine insecticides in England and Wales. (*a*) dieldrin sheep-dips, 1961–63; (*b*) dieldrin, aldrin, and heptachlor used in arable crops, 1962–63; (*c*) all organochlorines used in arable crops and orchards, 1962–63.

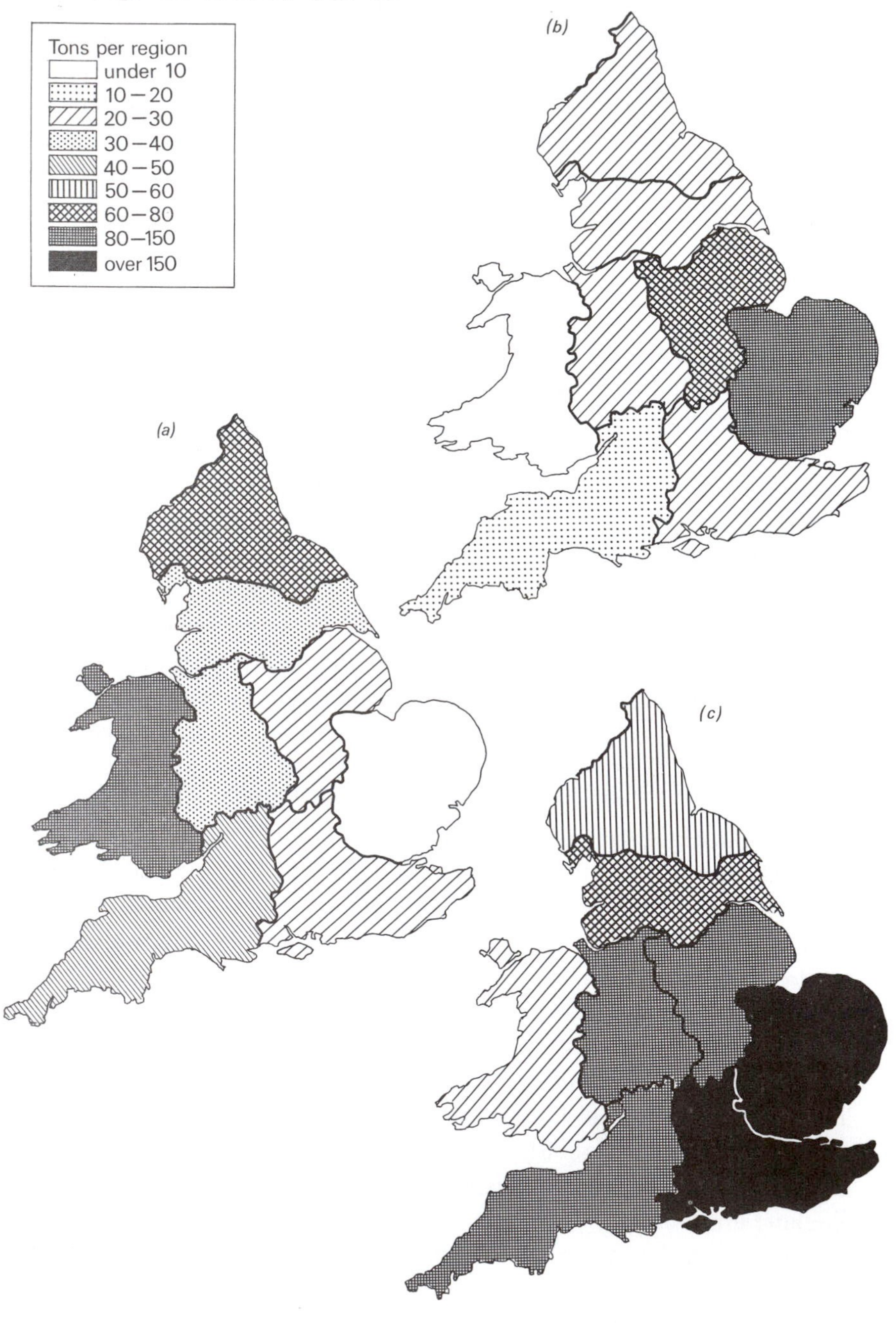

breeding success of raptors which take carrion, such as Golden Eagles and Red Kites, declined but Parslow (1973) reported little change in populations, as did Prestt (1965) for the Buzzard. Interestingly, however, the map for Sparrowhawk given by Prestt shows the area of greatest change stretching across midland Britain from East Anglia to North Wales and thus embracing the regions of maximum organochlorine usage in cropping and in sheep farming. Nevertheless the main impact of organochlorine compounds on birds was reported from the arable areas of the east rather than from the pastoral west, indicating that their use as crop chemicals was much more harmful to birds. This perhaps reflects the extent to which birds exploit arable farm operations and waste for food.

The decline of the Peregrine is the classic example of the effects of organochlorine pesticide contamination (Ratcliffe, 1980). When a national survey of Peregrine numbers was conducted by the BTO in 1961 (Ratcliffe, 1963) the population had already fallen to only 68 per cent of its pre-war level and to 44 per cent by 1963. The decrease was greatest in arable areas, lower where permanent grass was dominant, and least where there was little or no agriculture. The beginning of the decrease in southern England and Wales coincided with the spread in use of dieldrin compounds between 1956 and 1958 and its intensification and spread northwards between 1959 and 1962 coincided with the greatest volume of use of these chemicals. Dressed seeds were eaten by pigeons and other prey species of the Peregrine and toxic residues concentrated along the food chain, eventually poisoning the falcons. DDT was also implicated, causing a marked reduction in eggshell thickness (in turn leading to greater egg losses from breakage) between 1956 and 1962. The decrease continued into 1964, when there were only three territories occupied in England and the Welsh population had fallen to 13 per cent of its pre-war level. But as a result of the voluntary ban on the use of persistent organochlorines on spring-sown seed, followed by further restrictions later, a recovery began. By the time of a repeat survey in 1971 the population had reached 54 per cent of its 1930s' level and by the time of a further survey in 1981, about 90 per cent (Ratcliffe, 1984). Although persistent organochlorine products have been withdrawn for most uses, a very few specialist uses remained until 1984, mainly in areas of intensive tillage. Where records are available, they show Peregrine (and Sparrowhawk) numbers are still depressed in these areas.

The Sparrowhawk was the most numerous raptor generally affected by organochlorines. Cramp (1963), Prestt (1965) and Parslow (1973) summarised the changes in status in the 1950s and 1960s, when a sharp decline was particularly evident throughout lowland England. Prestt (1965) noted that by 1963, the end of his survey period, 'From being one of the commonest and most widely distributed diurnal birds of prey, there is not now a single county remaining in England where the species can be considered a common breeding bird'. Newton & Haas (1984) noted that the main decline occurred in the period 1957–63, following the introduction of the cyclodienes (dieldrin, aldrin, and heptachlor) into agriculture and that the greatest declines coincided with the greatest areas of tilled land. They pointed out that even the 60 per cent fall in breeding success recorded since 1947 as a result of DDT usage (Newton, 1974) would not have been sufficient to cause a decline of the speed and scale that occurred, which was in any case associated with many dead Sparrowhawks being found. Newton & Haas therefore argued that adult survival decreased with the advent of the cyclodienes and that the marked recovery that has taken place (Fig. 9.6) has been based on improved adult survival following the progressive withdrawal of these chemicals. Moreover, although breeding success has remained depressed, Sparrowhawks spread

Fig. 9.6. The recovery of the Sparrowhawk population of farmland in Britain, 1962–84, as shown by the percentage of CBC farms with the species present (open circles) and by the national CBC index of territorial birds for all habitats (dots). CBC index for 1974 arbitrarily set to 100. Data from Marchant (1980, and personal communication).

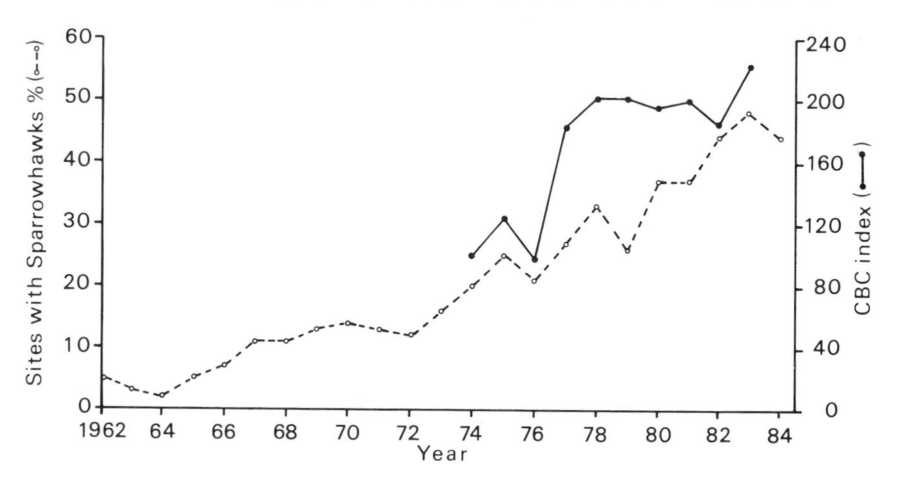

west to east across the country in a wave pattern in the wake of these progressive withdrawals, starting in the least tilled areas.

Nevertheless, some areas had recorded extensive declines before the introduction of the cyclodienes. Thus in Kent the species had apparently ceased to breed by 1953 (Taylor *et al.*, 1981). Shrubb (1985a) examined the impact of organochlorines on Sparrowhawks in Sussex and Kent in detail. Here recovery occurred in two clearly defined stages, with a rapid partial recovery from the very low point reached about 1960 until about 1967, following a 92 per cent reduction in cyclodiene usage after restrictions introduced from about 1962. Even so, by 1970–74 the population was still under 50 per cent – and perhaps as few as 15 per cent actually nesting – of its probable potential level. Breeding performance, defined as the ratio of proved breeding records to occupied sites, continued to decline during this period, reaching its lowest point in 1973. The second, and main, phase of recovery was associated with greatly improved breeding performance from about 1976, which included a significant increase in the number of broods of three or more young being reared, and which followed a 75 per cent reduction in DDT use, mainly after 1969 in orchards. Despite the improved percentage of pairs actually breeding, the breeding success of the species as a whole remained depressed throughout the period 1960–84, as it did nationally.

This study also confirmed the importance of drift in dispersing such chemicals through the environment, a point stressed by Newton & Haas (1984). Except for the extreme east of the county, where orchards are abundant, East Sussex has more woodland and less cultivated land than any comparable area of the English lowlands. It should, therefore, have formed a reservoir of breeding Sparrowhawks as, indeed, did the downland forests of West Sussex which are islands in a sea of cereals. Instead, no breeding was reported for the major part of East Sussex for 1960–75 inclusive. Westward drift from orchards in Kent and extreme East Sussex was the probable cause, for examination of the weather records for the period indicated a high probability of easterly winds during the times spraying would be done. Similar effects are probably responsible for the Sparrowhawk's persistent failure to recolonise much of the primary tilled area of eastern England, where there has been similarly high DDT usage in a concentrated area, in this case of vegetables (especially brassicas) rather than of orchards.

The decline in organochlorine usage shown in Tables 9.4 and 9.5 does not include contemporary changes in the toxicity of the chemicals remaining in use. In 1965 the use of cyclodienes in cereals was restricted to wheat-bulb fly areas and they were withdrawn completely from cereal growing in 1975. Dieldrin was also used throughout the sugarbeet area until 1975 and remained on the ACAS Approved List – as a dip for cabbage plants – until 1980. As a sheep-dip it was withdrawn in January 1966 and largely replaced with organophosphorus compounds. Although the main reduction in cyclodiene use occurred in the 1960s and early 1970s, the pattern of declining toxicity has continued. HCH, the least toxic organochlorine compound in regular use, occupied nearly all the spray area by 1981–82 and comprised nearly 80 per cent of the total tonnage of organochlorines used then, compared to about 60 per cent in 1966–70.

Many small seed-eating birds died in large numbers in the years 1956–60, due largely to the use of cyclodienes as seed dressings (Cramp & Conder, 1961; Cramp, Conder & Ash, 1962), but incidents of mass mortality ceased soon after their withdrawal from spring use in 1962. Sublethal effects of organochlorines, e.g. due to eggshell thinning, have been widely reported of large species. Raptorial species have been most affected (see above) but Rook and Carrion Crow have each experienced average shell thickness reductions of about 5 per cent (Newton, 1979;

Table 9.6. *Changes in farmland nest success[a] in various species in Britain, 1962–80*

	Percentage of nests successful			
	1962–65	1966–70	1971–75	1976–80
Skylark	26.8	28.0	36.6	47.1
Pied Wagtail	56.0	80.6	58.9	75.9
Dunnock	32.8	23.1	45.6	47.6
Blackbird	29.4	47.3	35.7	48.2
Song Thrush	36.4	39.7	43.7	46.8
Mistle Thrush	36.8	41.9	50.5	49.4
Whitethroat	62.3	71.7	84.0	65.4
Magpie	37.7	14.9	32.3	36.0
Carrion Crow	15.8	34.4	45.8	39.6
Starling	54.2	70.9	69.9	77.9
Chaffinch	41.2	27.3	41.4	34.5
Greenfinch	30.4	39.7	46.6	38.1
Goldfinch	36.2	53.8	50.8	53.7
Linnet	42.6	34.4	43.1	34.5
Yellowhammer	27.9	27.8	24.6	38.9
Reed Bunting	35.7	50.0	51.9	55.6

[a] Percentage of nests fledging at least one young.

Ratcliffe, 1980). Smaller birds are less frequently considered in this context but Table 9.6 shows that nesting success of many small passerines has in fact increased since the early 1960s. The use of nest success here provides only the crudest of measures (see Appendix 2) and in some cases results here are compounded with changes in management (e.g. Linnet). A few species show recent decreases in success, e.g. Whitethroat, Carrion Crow, but the general picture is of improved nesting success. For 11 of the 16 species considered, 1962–65 was the worst period of the four; 1966–70 was the worst period for only four species; and 1971–75 was the worst for only one species. On the other hand, 1976–80 was the best period for seven of the 16 species. Detailed examination of breeding success in a variety of regional samples of various chemical history suggests that improvements in hatching success are responsible. This is as expected if pesticides resulted in generally reduced nest success in the early 1960s, but no doubt other explanations could be advanced. Of these the most worrying is the risk that average measured success has increased because farmland birds now never breed in 'poor' environments. If such were the case, the Nest Records Scheme would not detect any reduction in nest success and might even show increased success because of fewer records from what used to be poor (but still acceptable) breeding habitats. Both Lapwing and Stock Dove have fitted this pattern at certain times (Murton, 1971; O'Connor & Mead, 1984).

The use of herbicides

While greater control of insect pests and of disease has progressively increased the yields of modern farming systems, it is the control of competitive weeds by herbicides which has been crucial to their establishment. Herbicides are used essentially in three stages of cultivations. First, they may be used in preliminary cultivations or cleaning operations, which may even include using glyphosate in the standing crop just before harvest, to clear crop residues and perennial weeds (particularly grasses). Second, they may be used as pre-emergent herbicides, applied at sowing. Herbicides of this type have been very valuable in root and winter cereal crops and kill weeds (especially grasses in cereals) as they germinate and grow through the chemical. Third, they may be applied in the early stages of crop growth as post-emergent herbicides. Some herbicides can be used either as pre-emergents or as post-emergents. Table 9.7 and 9.8 summarise the growth and development of herbicide usage in selected crops since 1955. It should be stressed that Table 9.7 is based on chemicals in the ACAS Approved List and not on the number of products.

Table 9.7. *Herbicides*[a] *available for weed control in selected arable crops 1955–85*

Year	Cereals				Roots		Peas/beans		Clearing[b]
	Not undersown	Undersown with grass or clover	Control of wild oats or grasses	Autumn sprays	Pre-emergent	Post-emergent	Pre-emergent	Post-emergent	Perennial weeds
1955	5	0	0	0	0	0	0	0	0
1960	9	3	0	0	1	0	1	2	2
1965	11	3	1	0	6	4	4	3	4
1970	15	4	5	3	18	9	7	5	4
1975	24	6	15	4	16	7	7	7	4
1980	26	8	21	9	14	12	9	8	5
1985	29	11	17	9	21	20	14	10	11

[a] Some chemicals have both pre- and post-emergent applications and are included under both headings.
[b] Chemicals used to clear perennial weeds from existing stubbles or prior to planting, irrespective of the crop to be planted.

Table 9.8. *Summary of herbicide usage in selected arable crops in England and Wales, 1955–85*

Pesticide survey cycle	Crop[a]	Area grown 000 ha	Area sprayed[b] 000 ha	Amount used tons AI[c]
1966–70	Cereals and other arable crops	4763	3643	6000
	Roots and vegetables	582	407	760
1971–74	Cereals	3246	4475	8727
	Vegetables	221	228	351
	Other arable and fodder crops	2083	1159	5864
1975–79	Cereals	3209	4576	7582
	Vegetables	250	395	1159
	Other arable and fodder crops	5804	2711	10619
1981–82	Cereals	3444	7395	12627
	Vegetables	195	852	1117
	Other crops	587	1544	2051

[a] Other arable and fodder crops include temporary grass, except in 1981–82, where 'Other crops' are defined as roots in Table 9.3.
[b] For definitions see Table 9.3.
[c] Active ingredient.

Probably the most significant feature in herbicide development in recent years has been the development of pre-emergent herbicides. These have the advantage of eliminating competitive weeds at crop establishment; 'tramlining' is particularly associated with their use in cereal farming and has led to a much more exact and accurate application, which is combined with very early applications of nitrogen in the spring to promote tillering. This combination greatly increases the capacity of the crop to smother weed growth and further use of herbicides is often unnecessary or is aimed at specific and difficult plants. Overall these techniques are steadily reducing the capacity of cereal field weeds to set viable seed. In general, what weeds there are in well-managed cereal fields tend to appear as the crop approaches maturity, when the lower leaves die off and allow more light to the soil. As a result of the rapid cleaning of stubbles and the autumn drilling now typical, little chance is given to this growth to seed. The use of pre-emergent herbicides in root crops is possibly even more serious for birds since roots are in principle better feeding areas than are cereals. Root crops are prone to infestation by plants such as fat hen or knotgrass, since their wide spacings at sowing allow more light to the soil. Such weeds could not be treated with the sprays first developed, for root crops were themselves vulnerable. Weeds therefore flourished each season until the crop plants were wide enough to cover the gaps. These weeds are important food plants for finches.

The decline in weed seed production is a change having serious implications for birds, as discussed for the Linnet above (Chapter 8) and for the Stock Dove below. Historically the weed flora of fields has largely been introduced with crop seed during sowing. Most arable weeds have adapted thoroughly to the patterns of cultivation and maintain themselves by seeding. They do not necessarily behave strictly as annuals, for the seeds of many can sustain long periods of dormancy and some, e.g. field poppies, clovers, wild oats, always throw a percentage of seeds which are naturally dormant and will not germinate until their period of dormancy is over. Thus the soil contains an extensive seed bank, built up over many years, and it is this that herbicides are slowly eroding, continuing more efficiently the process started by drill husbandry.

Table 9.9 summarises the capacity for dormancy and seed production of some common arable field weeds which are also listed by Newton (1972) as important food plants for finches. The figures for seed production should not be taken too literally but are simply a guide to the productive capacity of these plants, based on experience of serious infestations in cereals of plants such as fat hen. The combination of a high productivity and a pronounced capacity to remain dormant but viable explains why few species have yet been eliminated as field weeds, despite 38 years of regular herbicide use. The decline in seed production, however, has virtually removed their once considerable value as a food source to seed-eating birds. Many are now uncommon enough to be fairly insignificant members of the farmland plant community, seeding only when weather conditions make crop management tasks difficult or impossible.

By their very nature most herbicides are selective and the susceptibility of weeds to the chemicals used also varies. Thus among annuals listed in

Table 9.9. *Dormancy and seed production in field weeds which are important finch food-plants*

Weed species	Food plant for	Periods of dormancy for viable seeds[a]	Seed production[b] g/ha
Chickweed *Stellaria*	Chaffinch Linnet Greenfinch Siskin Bullfinch Goldfinch	10 years 22%	74710
Groundsel	Greenfinch		6523[c]
Charlock	Chaffinch Linnet Greenfinch Bullfinch	many years	299000
Persicaria	Chaffinch Linnet Greenfinch	21 years 55% 30 years 9%	81500
Knotgrass	Chaffinch Linnet Greenfinch	60 years	179000
Thistle			
Cirsium	Greenfinch Linnet Goldfinch Siskin	21 years 5%	21000
Carduus	Greenfinch Linnet Goldfinch Siskin		592800
Sowthistle			
annual	Goldfinch Bullfinch	10 years	107500
perennial	Goldfinch Bullfinch	6 years 40%	19000
Fat hen	Chaffinch Linnet Bullfinch	21 years 66% 30 years 9%	143250
Spurrey	Chaffinch	5 years	18000
Docks	Bullfinch Siskin	39 years 5–6%	269230
Chickweed *Cerastium*	Chaffinch	40 years	54340
Buttercup	Linnet Bullfinch	many years	173000

[a] Percentage viable stated where known.
[b] Assuming 5 plants/m^2.
[c] Seeds per plant.
Based on Salisbury (1961) and Newton (1972).

Table 9.9, charlock and fat hen are easier to control than chickweed, knotgrass or persicaria, while perennial thistles are more readily controlled than perennial sowthistles; grasses are more difficult to control than dicotyledonous weeds. Such selective elimination of weeds can result in species difficult to control becoming dominant elements of the field flora and exacerbating the problems of field hygiene. This at least partly accounts for the continuing increase in herbicide use.

The Stock Dove – a case history of chemical effects

The fortunes of the Stock Dove in Britain largely reflect the successive impact of seed dressings and herbicides (O'Connor & Mead 1982, 1984). Its population fell steeply between 1951 and 1961 (Fig. 9.7), because its dependence on seeds, and particularly its penchant for feeding on weed seeds and newly sown grain on ploughed land, made it rather vulnerable to poisoning by organochlorine seed-dressings. Following the withdrawal of the cyclodienes from 1962 the population has partly recovered but has done so more slowly than it collapsed. The main factor behind the population changes has probably been change in breeding success. This fell from 14.7 per cent in 1942–49 (before the use of organochlorine seed-dressings) to 7.5 per cent in 1950–59 and 9.0 per cent in 1960–69,

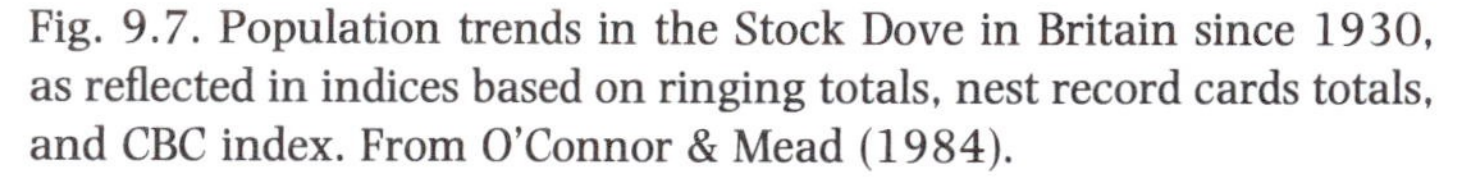

Fig. 9.7. Population trends in the Stock Dove in Britain since 1930, as reflected in indices based on ringing totals, nest record cards totals, and CBC index. From O'Connor & Mead (1984).

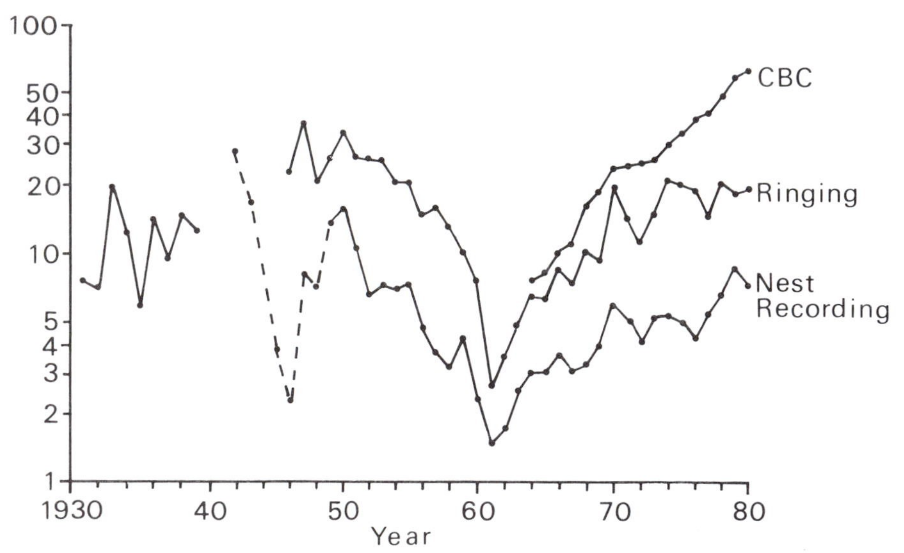

before recovering to 14.4 per cent in 1980. The changes in breeding success are actually more pronounced than these figures suggest, for the worst hit habitats (e.g. arable land) and regions (e.g. the Midlands) were largely abandoned in the 1950s. As a result, the estimates of nest success for that period are boosted by the paucity of nest records from the habitats poorest in success at that time (O'Connor & Mead, 1984).

Despite the withdrawal of cyclodienes, the recovery is incomplete and conditions for breeding of Stock Doves have not returned to those of the 1940s. These points seem largely due to the impact of management changes on food supplies. Three changes in particular may be important. First, winter fallows and old stubbles are the main sources of weed seeds for overwintering birds and fewer of these survive the autumn since the increase in autumn-sown cereals (p. 59). Second, weed seeds have declined in abundance due to herbicide use, as discussed above; of particular relevance here has been the development since the early 1970s of chemicals specific to the *Polygonum* group, since this has directly hit the main food plant of Stock Doves. Finally, with 'tramlining' techniques, cereal plantings are established and maintained at more even densities, effectively removing the bare, thinly vegetated patches on which Stock Doves prefer to feed. Thus the cereals environment particularly is now less favourable to this bird than before 1945.

This emerges most clearly from regional analysis of population levels and breeding success. In southern England, where the recovery is as yet incomplete, daily egg mortality has risen from 4.9 per cent in the 1950s to 8.7 in 1980 (though partly offset by nestling mortality falling from 5.5 per cent to 3.7 per cent over the same period). Similarly, in eastern England, where the recovery has levelled off, egg mortality rose from 4.8 per cent to 5.9 per cent (again offset by nestling mortality moving from 6.9 per cent to 6.2 per cent). In western England, in contrast, both egg and nestling mortalities have fallen and the population has continued to rise there. Seasonal variation in breeding success has also altered in recent years. Nestling success used to be best from second clutches, normally laid between May and July. Murton (1966) emphasised the importance to Stock Doves of weed seeds at this time. Comparison of Stock Dove breeding seasons in the 1960s and 1970s shows that this June peak of laying has now largely disappeared (O'Connor & Mead, 1982). This change is partly correlated with the greatly reduced cultivation of root crops, turnips and potatoes, which provide the normal sources of open ground within which to seek weed seeds. More important, though, is the greater cleanliness of these crops where they are still grown. The use of herbicides (Table 9.8) now ensures that such crops carry lower weed populations than in the past, to the detriment of the Stock Dove. In this the Stock Dove resembles the Linnet discussed in Chapter 8.

Pesticide control and farmer attitudes

With pesticides it seems probable that lethal direct effects are now fairly well under control, not only through the Pesticides Safety Precautions Scheme but also through the monitoring work by MAFF and DAFS. But Stanley & Hardy (1984) also stress the importance of understanding the ecological implications of chemical use discussed by Potts & Vickerman (1974). Their paper was particularly concerned with the relationship between cereal production, pesticide use, and the entire farming system, and whether reliance on pesticides could promote further pest, disease, and weed problems. The question of pesticide use creating further pesticide use – the so-called pesticides treadmill – has general implications which we are only just beginning to understand (p. 198). Stanley & Hardy identify this as a major research priority. Since pesticides are the key to modern farming, and since their ecological implications for birds are virtually unstudied, we would not disagree.

Although the control mechanism of the Pesticides Safety Precautions Scheme has proved successful in limiting the direct effects of pesticides, the attitudes of farmers to their use and the adequacy of advice given to farmers need consideration. These points have been considered by Tait (1977), particularly in relation to the use of DDT in brassicas. Tait conducted a survey of farmer attitudes towards the various risks – financial, health, and environmental – involved in pesticide use. She found that, although most farmers favoured pesticide use on financial grounds and were averse to pesticide use on health and environmental grounds, only the financial argument had any influence on the actual use each farmer made of pesticides. This applies particularly to the insurance use of pesticides, where chemicals are applied even though no evidence of the pest is present, rather than risk a subsequent loss. Tait found that the 'official' position in favour of a voluntary system regulating pesticide use did not work where brassicas and other crops are grown intensively in certain regions of Britain. The ACAS Approved List contains a number of internal contradictions as to the status of certain chemicals and this, coupled with the aversion to financial risk, resulted in most farmers using the chemicals in the context most favourable to their own position. The perception of the risk to wildlife also varied with farmer attitudes to wildlife. In Lincolnshire, where hedges, scrub, and related habitat have long been scarce, most farmers were aware of wildlife but only as 'vermin'. In Norfolk and Suffolk, where the distribution of hedges is patchy, many farmers were unaware of wildlife. In Bedfordshire, where wildlife habitat has long been abundant, most farmers were both aware of and in favour of wildlife. Tait concluded that advice given to farmers rarely included adequate information on environmental issues, a shortcoming reflected in regional hostility to wildlife of all forms because of misunderstanding of wildlife relationships to farm crops.

Fertilisers

Pesticides do not provide the only source of environmental pollution arising from farming. Problems can also be created by intensively managed stock and by high levels of fertiliser usage, both of nitrogen and phosphates. Nitrogen in either granular or liquid form is by far the most widely used manufactured fertiliser, especially in intensively managed grassland areas, and is the most likely potential source of problems. Table 9.10 illustrates the growth of nitrogen usage during the period 1962–82, both in terms of tonnage used (as kg/ha) and as percentage of crop dressed. The table shows that the greatest changes have occurred in grassland, other than rough grazings. The area of pasture dressed and the dose rate applied to all improved grass have been the major changes. In tilled land the increase in dose rates since 1974 derives largely from cereal growing, particularly the switch to winter wheats. The very high level of nitrogen used on grass reflects the high stocking rates now general. Some regional variations were discussed in Chapter 5, under general grassland management, but differences in the scale of change shown in the table suggest quite clearly that such regional variations are slowly disappearing. There is, however, a striking difference in the patterns of use of nitrogen and of the other main plant foods used, for in the period 1962–82 the increased dose rates shown in Table 9.10 were peculiar to nitrogen.

It is clear, therefore, that pollution problems are most likely to arise from the use of nitrogenous fertilisers. Such problems may be caused by several factors. They may be indirect, as high fertiliser rates allow high stocking rates on small areas, which can produce problems in disposing of animal effluents, themselves a pollutant. Secondly, the present methods of using nitrogenous fertilisers are wasteful, in the sense that the crop plants do not always taken them all up. Inevitably some nitrogen leaches into drainage systems, particularly as good and rapid drainage is an essential prerequisite of intensive farming. Both factors can have a deleterious effect on surface water. Nitrate levels are also rising in underground aquifers, a matter of

Table 9.10. *The growth of nitrogen usage in England and Wales, 1962–82*

	Tilled land		Leys		Permanent grass		Total	
	% area treated	kg per ha[a]	% area treated	kg per ha[a]	% area treated	kg per ha[a]	% area treated	kg per ha[a]
1962	88	84	85	42	37	20	85	44
1966	84	73	69	67	44	31	59	55
1970	90	88	84	107	57	51	77	78
1974	92	85	86	133	63	69	81	89
1978	—	104	—	149	—	87	—	107
1982	96	141	93	186	77	96	89	132
Change-over period %	48	120	43	343	108	380	24	200

[a] These figures are kg nitrates per ha of crop grown.

great concern in East Anglia today. Thirdly, in old pasture, continuous nitrogen applications markedly alter the plant community: different grasses and herbs differ in their reactions with nitrogen and the sward becomes dominated by species best able to use it. More importantly, old pastures rarely use nitrogen effectively and are therefore not treated with it. Hence high nitrogen usage in pasture is a good index of the extent of ploughing and reseeding with high output rye-grass swards. This represents a major habitat change whether or not it is accompanied by improved drainage.

Data on the impact of inorganic fertilisers on soil invertebrates are sparse, although Edwards (1984) recorded that a long-term experiment over 118 years at Rothamsted showed an inverse correlation between nitrogen levels and populations of soil invertebrates. In this experiment, though, the highest level of nitrogen used was 40 per cent below the average used in intensively managed grassland today.

Summary

Pesticide use in Britain has grown steadily since 1951 and has been a major instrument in producing changes in cropping patterns and rotations. Since the 1970s the use of fungicides and insecticides in cereal farming has greatly increased. This, together with use of more sophisticated herbicides, has impoverished the cereal environment as a habitat and has been reflected in changes in Grey Partridge and Stock Dove populations. Although the diversity of arable weeds has remained unchanged, the general decline in the abundance of weed seeds as a result of herbicide use has reduced their proportion in the diet of birds. The implications of a possible pesticides 'treadmill' need thorough examination. Fertilisers, particularly nitrogenous compounds, are also a potential source of pollution, though data on the ecological implications of this are extremely sparse.

10: Crop protection, shooting and persecution

Rather few bird species fall into the pest category, those that do being mostly large, social or gregarious species, which are either herbivorous or granivorous in diet. Pest species are, in general, pre-adapted to exploit some part of the man-made environment and in periods of rapidly changing land management we can therefore expect new agricultural regimes to be open to the flexibility of old-established pests or to create openings for new ones (Flegg, 1980). Pest problems may arise through features of the habitat and its environment or through features of the pest species itself. Wiens & Johnston (1977) have shown that most bird pests share a number of ecologically important traits that predispose them to successful exploitation of agricultural crops. Bird problems are therefore particularly likely to arise where vulnerable crops are grown in close association with concentrations of birds. Bullfinch attacks on fruit trees in Kent orchards, for example, are more severe where the orchard has been planted adjacent to natural woodland and the pattern of attack is consistently of orchard rows adjacent to the wood being the first damaged (Flegg, 1980). Similarly, the sudden outbreak in 1973 of Pink-footed Goose damage to carrot crops in

Lancashire may have been stimulated by the local practice of leaving carrots in large piles around fields which were initially grazed harmlessly by the geese.

Bird problems have no perceptible impact on farming in terms of national food production (e.g. Feare, 1980); however, they can be serious to the individual farmer. The actual value of damage done (or claimed as done) by birds to crops is often very difficult to assess; indeed it is striking how few farmers are prepared to assess this with any accuracy. Damage is usually fairly minimal and cases in which a crop is a total loss are rare. Damage costs derive from two sources, the actual loss of yield and the increased inputs needed to repair any damage. In many circumstances the reasons for yield losses are only partly attributable to birds and separating the causes is difficult. Woodpigeons, for example, tend to select weaker areas of a crop because the plants are easier to reach. Standards of management may exacerbate problems, or even create them. This may well have happened, for example, in upland stock rearing, particularly in Wales, where sheep numbers rose by 49 per cent between 1963 and 1983 while the area of sheepwalk available has been progressively restricted by afforestation. Inevitably there is more sheep carrion to encourage the corvid population (Chapter 8).

Damage to crops

Herbivorous species affect agriculture mostly by the grazing of young growing crops, or by pulling or digging up young plants in row crops such as sugar beet. Grazing is a problem mainly in cereals, young grass leys, brassicas, clovers and other leguminous crops. In cereals geese and swans can cause two problems, direct grazing and 'puddling' with their big, flat feet (which caps the soil and restricts root development). The same species may also damage young leys by hard grazing in winter. Moorhens will very occasionally graze cereals extensively and wildfowl such as Wigeon also graze such sites. Problems in brassicas, clovers and other legumes are largely caused by pigeons, although geese will graze brassicas in winter. In each case the grazing checks growth rather than destroys the crop but the crop has to use energy to replace that growth. In cereals this in turn affects the number of tillers and the size of the ear, so yield is generally reduced. Even where grazing stimulates later regrowth, the timing of attack may be important. In grassland in Scotland Pink-footed and Greylag Geese grazing the first flush of grass may deprive farmers of the important 'spring bite' prepared for farm stock, and later regrowth does not compensate for this loss.

The pulling or digging up of row crops may be to eat the plants themselves or to find prey animals underneath. This damage differs from that done by grazing, in that the uprooted plants die. Skylarks, which commonly eat sugarbeet seedlings, and Rooks, which will incidentally dig up plants while foraging around them, are perhaps the species typically concerned. Rooks were at one time a serious problem in sugarbeet, because they dug up plants while looking for wireworms around the plant roots. The habit has disappeared following the use of HCH and of granular carbamates to reduce the wireworm populations (Dunning, 1974). The problem of Skylarks seems to be a new one and has probably arisen only since root crops have been drilled to a stand, rather than thinned to a stand by hand-hoeing after emergence. This is another example of the point made by Flegg (1980) on p. 216.

Granivorous species may dig up seed or may take grain in the ear. Many species take seed left on the surface after drilling, particularly in cereals and in peas and similar crops. As this is wasted seed, its removal hardly damages the crop. Starlings and Rooks, and less frequently other corvids, dig seed up. Rooks are especially fond of oats, which they husk before eating, and Rook attacks on spring-sown barley and oats in north-east Scotland have considerably reduced the crop yields (Feare, 1974). Starlings usually attack winter wheat and barley in November and December, mainly in late-sown fields soon after the plants emerge above ground, though attack is sometimes earlier. Seeds are taken either directly from the soil, leaving an apparently intact plant, or by uprooting the plant before taking the seed, and attacks are most prevalent in fields close to the winter roosts where many thousands of birds may gather and feed briefly before entering the roost itself. This sort of damage was widespread in the 1920s (Collinge, 1924–27) but then declined until the mid-1970s. The main factor behind the recent increase appears to have been the spread of shallow drilling, a practice designed to promote more rapid plant establishment and to combat wheat-bulb fly. However, this leaves proportionately more of the seed vulnerable to Starlings, which can penetrate only the top 50 mm or so of the soil (Boyce, 1979).

All granivorous birds will take grain from standing cereals, however, and finches are especially likely to take seed from standing oilseed rape crops, which may now be an important summer food for them, and pigeons in particular will raid pea and bean crops. Only sparrows and Greenfinches appear readily able to feed on cereals which are standing erect, although most finches can cope with oilseed rape in this way. Corvids, pigeons and gamebirds need either laid patches, where they can in effect feed from the ground, or access such as an open side from which they can walk into the crop, treading down the stems to get at the ear. Standing crops are otherwise often immune to their attentions. This may be an important factor in considering changes in Rook numbers (see Chapter 8), though Feare (1982) provides an interesting photograph of Rook damage to winter wheat along the line of a fence. The birds would jump from the fence onto the standing wheat to bring the ears to the ground. Such damage to winter barley is fairly common in Hampshire and Wiltshire (G. R. Potts, personal communication).

Bird pests also affect two other areas of agriculture: orchards and indoor stock rearing. The Bullfinch is the main culprit in damage to orchards, particularly to fruit buds, but the Starling has also caused extensive damage in cherry orchards, taking up 21 per cent of the crop where too few fruit pickers were available to harvest the fruit as it ripened (Feare, 1980). Stock rearing suffers because many species take advantage of the food supplies provided by the management of farm animals, taking food from stock-yards or food stores (see Chapter 4), though probably only Collared Dove, Starling, House Sparrow and sometimes corvids are ever sufficiently numerous and tolerant of man to provide a serious nuisance. It is important to distinguish between this kind of exploitation and the use a wide range of small passerines make of stock-yards in winter, where they scavenge seeds and insects from hay, bedding and dung (Chapter 4). The quantity of feed grain taken from food stores and the number of individuals regularly present can be unexpectedly large. Thus a flock of 800 Collared Doves living around a large dairy in south-west Sussex was estimated to take as much as 50 kg of grain a day, equivalent to 18.25 tonnes a year; corvids can cause similar losses. The removal of animal foodstuffs by Starlings has been found to be economically serious both at poultry farms and in cattle-rearing units (Murton & Westwood, 1976; Wright, 1973; Feare, 1980). Starlings may take up to 12 per cent of the food presented to calves (Feare & Swannack, 1978) and calves reared in pens protected from birds are able to grow faster and convert food more efficiently than are calves in open pens (Wright, 1973). Immigrant birds rather than residents, and dominant males rather than females, were more numerous at the feeding troughs. Similarly, in dairy herds losses of feed barley from the cow ration amounted to 10 per cent over four months and may be associated with claimed reductions in milk yields. The fouling of foodstuffs may also make them less palatable to the cattle.

Environmental and management factors

The account above gives the impression of constant conflict, but there are limitations to the extent any birds can damage crops, quite apart from those imposed by preventative measures taken by the farmer. The period during which crops are vulnerable is usually limited. Crops such as cereals, oilseed rape or clover are normally attractive to grazing birds only during the early stages of growth, so that damage is restricted to winter and/or early spring in most cases. Then only certain fields are selected – not every rape field is preyed upon by pigeons nor are all cereal fields near roosts raided by ravening hordes of geese! Selection is strongly influenced by such factors as field size, the nature of the boundaries (geese will rarely feel safe in a small, well-hedged field), and by the closeness of buildings, roads or regularly used paths or tracks. Probably the large open fields typical of modern arable farming are more vulnerable than the smaller fields of the past. Traditional patterns of behaviour may also affect choice. Damage by some species is limited by their distribution – geese, for example, are confined to the vicinity of estuaries or large inland waters, used as roosts. How restricted are the problems they create is well illustrated by the Brent Goose, which affects fewer than one per cent of farms in Essex and West Sussex, the counties which have reported most problems. Problems with Barnacle Geese in areas such as Islay are similar in scale. With up to 20000 birds concentrated on rather few farms, losses are important to the individual farmer but not to agriculture as a whole.

The problems caused by birds stealing animal feed from intensive units differ from those in fields where there are natural limitations to the extent of damage, particularly in the period of time involved. In buildings the problem exists all the time, if it exists at all. Furthermore, adequate ventilation is essential for stock housed in asbestos-clad buildings, which makes the restriction of access to birds more difficult. Possibly more serious problems arise where the animals themselves are restricted in movement.

Bird damage to agriculture frequently occurs where alternative resources are not available to the birds (Newton, 1964b; Feare, 1974, 1978; Feare *et al.*, 1974; Dyer & Ward, 1977). Thus Newton (1968, 1972) argued that the pattern of heavy bud damage by Bullfinches to orchards in alternate years was the result of alternating good and bad annual production of seed by ash (*Fraxinus excelsior*), the preferred food. In poor years the ash seeds were exhausted early and the birds were forced to resort to the fruit buds. Newton showed experimentally that Bullfinches fed exclusively on a diet of buds were unable to maintain their body condition, thus indicating that the buds were a food of last resort. A similar problem has arisen with Linnets (Flegg, 1980). In orchards herbicides are increasingly used to keep the soil around the base of each fruit tree free of vegetation, and a combination of mowing and herbicide use keeps the tracks between the rows of trees free of broadleaved weeds. Similar 'clean' cropping is now the norm with soft fruits. The resulting shortage of natural weed seeds has driven Linnets to make greater use of the seeds in strawberry fruits, thus introducing a new pest problem! Indeed, in some areas Linnets now raid the strawberries under polythene tunnel cloches within which the warmer microclimate not only provides an early crop at a time of natural food scarcity but also considerable protection from the elements and from any deterrent devices the growers might use!

Farming has restricted natural food sources in other ways. Geese, for example, have lost many areas of coastal grazing marshes, which they would once have used without conflicting with agriculture. In such circumstances birds have adapted to the food resources provided by crops. Indeed Flegg (1980) has suggested that farmland management may be in effect forcing or encouraging some species to attack agricultural crops. The cases of Starlings, Bullfinches and Linnets quoted above are each an example of this. In such cases, if the value of any damage they do is comparatively low it is nearly always cheaper to accept it and to do what is possible to limit it by simple scaring techniques than to try to prevent it entirely. The latter is usually impossible anyway, since the basic distribution and numbers of the bird are entirely outside the farmer's control.

Preventing damage

Given the extent of pre-adaptations of pest species to the use of agricultural land, the most ecologically sensible approach to crop protection is to regard a bird pest as yet another environmental factor to be taken into account in considering modifications of existing agricultural practices or the introduction of new ones. If the value of damage is high it is almost certainly most economic to try and provide an alternative food source, for example, in refuge areas to which the birds are attracted in preference to the vulnerable crops. Trials of this technique have been conducted in Britain, based on considerable experience in the United States (Peterson & Fisher, 1956; Owen, 1980). Such refuges in the United States are farmed primarily for geese, to provide winter forage (cereals) and spring grazing (grass and clover). In some cases normal farm management can continue, so that the cultivated crops are effectively shared between the geese and farmer on an agreed financial basis. The availability of feed crops within the refuge is an important component in reducing conflict between birds and local farmers, for the mere protection of roosting sites concentrates the birds at night, only to disperse them onto adjacent agricultural land the following morning! A further risk of the refuge system is that geese will be encouraged to switch to agricultural crops in preference to their natural foods and careful assessment of the local conditions before the refuge solution is adopted is therefore essential (Owen, 1977). Again, if winter food is already limiting, the provision of food in refuges may result in an increased goose population. Refuges have been tried experimentally with Brent Geese in Sussex and Essex, by decoying the birds off cereals onto permanent pasture. Some 21 per cent of the entire Chichester Harbour population of Brent Geese used the RSPB sanctuary area on Thorney Island in the winter of 1982–83, when particularly large numbers were present and feeding outside the estuary (Prater, 1982).

Scaring is much the most widely used alternative to shooting, at least in field crops. The aim of such exercises is not to control the numbers of a pest but to limit its access to the crop. Scaring is based on four methods – placing scarecrows; constant disturbance (which is most effective if one man simply walks conspicuously across the area); using brightly coloured moving objects (blue fertiliser bags on poles and 'flash-harries' are widely favoured); and assorted 'bangers'. If carried out efficiently, scaring is usually all that is necessary to protect crops from significant damage. It is important with many species to prevent a pattern of behaviour forming, particularly with geese where behaviour patterns may be remembered in successive seasons. Scaring devices which rely on simulating shooting usually work best if they are moved about and occasionally mixed with real shooting. Many birds become accustomed to ignoring harmless bangs which are regularly timed in one spot. Effective scaring thus also includes anticipating problems by observing which fields are more favoured (and therefore vulnerable) and discouraging birds beginning to use them.

Preventing access is mainly needed in such operations as intensively managed livestock and other more specialised enterprises. Shooting, violent scaring, and poisonous or narcotic baits are often not practical around buildings where animals are kept and may in any event require prior licensing by MAFF or NCC. Restricting avian access to food stores or, less frequently, to feeding areas, is therefore the most feasible way of controlling such losses. Feare (1980) concluded that it was so difficult to identify and deal with the segment of a Starling population actually responsible for food losses at intensive calf-rearing and dairy units that effective population control to regulate the damage was an unrealistic goal. Instead, he advocated the modification of the agricultural habitat to make it less attractive to birds. Thus, in the case of the calf units the use of heavy duty PVC strips to exclude Starlings was a cost-effective solution (Feare & Swannack, 1978), whilst indoor housing in bird-proof buildings was the most efficient way of winning all the advantages of the 'complete diet' scheme. The substitution of less expensive waste grain from breweries for the crushed barley component of the cattle diet provides an alternative pathway to reducing the costs of losses to Starlings. Almost every case of bird damage in intensive systems has to be dealt with individually, as the layout and management of the system pose different problems in each. Careful observation of how birds are exploiting the opportunities, however, is a basic requirement for successful prevention.

Compensation and insurance schemes are two other methods of reconciling conflicts between agriculture and protected species of waterfowl. Their relative merits have been reviewed, in the light of Canadian experience, by Boyd (1980). The Saskatchewan Government used an insurance scheme from 1953 to 1977 to compensate farmers for waterfowl damage. Each farmer paid a two per cent premium to the insurance fund, in turn supplemented by a charge on the provincial gamebird licences, thus providing the means to compensate farmers experiencing significant damage that year. Since much of the damage caused by ducks was scattered and seldom recurrent on the same farm in successive years, the scheme proved economical in practice. When damage compensation schemes were tried in Canada, both in Saskatchewan in 1978 and in other provinces, the costs were considerably higher than for insurance schemes, and claims were rising to meet and exceed the budgeted funds. No less than 55 per cent of compensation claims came from townships that had not filed insurance claims in the 13 years preceding the introduction of the compensation scheme. In this the Canadian scheme resembles the situation in Britain over claims for compensation where agricultural development proposals in respect of Sites of Special Scientific Interest are blocked under the Wildlife and Countryside Act. Boyd points out that insurance schemes are preferable to compensation schemes because they can readily encourage changes in land use. Thus the Alberta Government uses a reducing scale of damage payments for each successive claim, thus providing incentive to cease growing crops in vulnerable areas. Another option is for the provincial or federal authorities to purchase large tracts of land in the areas most vulnerable to waterfowl damage.

The use of repellants to control bird damage has received some attention (review by Wright, 1980). Skylarks foraging in sugarbeet fields in spring, for example, prefer beet seedlings to the scarcer weed seedlings. Where the crop was treated with aldicarb, however, this preference was less pronounced (Green, 1980). In the United States methiocarb is widely used as a repellant on sprouting corn and on cherry crops. Repellants can protect crops successfully only if alternative foods are available, for otherwise the birds are forced to use the treated crop despite their aversion to the repellant (Rogers, 1980). Other ecological and behavioural constraints may likewise limit their effectiveness. Indeed, the inability to reproduce the results of experiments in space and time has been a major problem in field trials of repellants (Wright, 1980).

Population control

Population control has obvious appeal to farmers suffering damage. It is normally carried out by shooting, usually over the crop, and is rarely successful as a means of limiting the population. Murton (1971) discussed this subject at length and concluded that such attempts almost invariably failed because the number of birds shot in the course of pest control was below the natural mortality of the species concerned. Shooting simply replaced other mortality factors, usually starvation, and the size of the population, and therefore its impact on crops, was unaffected. In the case of Woodpigeon, for example, Murton (1965, 1968) provided evidence that the early removal of birds from the population merely allowed the survivors to feed and survive better. He also noted an in-built tendency to stop shooting just at the point where a measure of control might be achieved since, at that point, few shots were presented to the gun. Murton acknowledges the value of tactical shooting at the site of the damage. Shooting then either succeeds in moving birds elsewhere (usually to some more tolerant farmer's fields), so protecting that crop, or it succeeds in reducing the amount of time the birds spend feeding on the crop, so reducing any damage. But it depends for success on the existence of an alternative feeding area and if the shooting disperses the birds over several other vulnerable farms, the problem is worsened.

A second problem in shooting birds for control of agricultural damage is that local population densities are often regulated by social interactions such as territorial behaviour or dominance hierarchies. In such systems access to the desired resource – here the farmer's crop – is regulated by the behaviour of the birds already on site. If these birds are shot, an ecological vacuum is created which is filled by other birds previously restricted to less favoured areas moving in, so the problem of crop damage continues. Flegg

(1980) estimated the distances over which some common farmland birds would move in this way and showed that some 27 per cent of Woodpigeons move 11 km or more in summer. Similarly 36 per cent of Collared Doves, 28 per cent of Skylarks and 24 per cent of Rooks typically moved this distance in summer. In winter distances moved were somewhat greater again, and Rooks, for example, concentrated into larger winter roosts. Thus these typical farmland species have considerable potential for movement to fill in any population lows brought about by shooting, thereby negating the effects of the shooting effort.

The use of poison or narcotic baits is theoretically a more efficient method of control by killing since it provides the means of removing a high percentage of entire flocks. For most people (including farmers) such wholesale methods of destruction are anathema, however, because it is difficult to find baits adequately selective to avoid the indiscriminate killing of non-target birds. Their use is therefore either very strictly controlled (as in the case of narcotics) or prohibited, though this does not prevent their use in some circumstances (see below). Several accidental poisoning incidents recently reported show how deadly such methods might be (p. 233).

The best studied species in relation to possible population control has been the Woodpigeon, for which a wide variety of crop plants are now important foods. The Woodpigeon was the subject of a cartridge subsidy scheme in Britain until Murton, Westwood & Isaacson (1974) showed that battue shooting in February merely substituted for other sources of mortality, principally starvation on a limited winter food supply. Shooting of this kind is therefore an ineffective means of regulating the population. Newton (1970) suggested that it might be possible to limit the population below its natural level by artificially reducing the food supply, for example, by ploughing stubbles immediately after harvest. The birds would thus have a shorter period on grain and would resort to clover earlier, thereby increasing the number dying of starvation. This measure has effectively been introduced with modern cereal management. Potts (1981b) provides a review of the recent fortunes of Woodpigeon in Britain, using the returns from the National Game Census for Woodpigeon (Fig. 10.1). Severe mortality during the 1962–63 winter was followed by a slow recovery between 1966 and 1970, since when numbers have fallen more or less steadily. These changes are not due to the loss of clover, for neither clover and oilseed rape acreages in spring nor grain stocks in autumn affected the total survival from one year to the next (see Fig. 10.1). On the other hand, computer modelling showed that the Woodpigeon population in

England is more susceptible to shooting than was the Cambridge population studied by Murton *et al.* (1974). Nationally there is in fact no compensation for losses attributable to shooting. Potts suggests that shooting pressure prevented Woodpigeons from recovering quickly from the 1962–63 winter. The withdrawal of subsidies in 1965 for battue shooting and by 1969 for other shooting therefore assisted the rise of the population in the late 1960's (Fig. 10.1). The decrease since 1973 may be associated with greater demand for Woodpigeons for the table. One must bear in mind here, though, that data presented in Chapter 8 show that Woodpigeon breeding parameters have also altered, a change that has not been taken into account in Potts' model, and that doing so would alter the amplitude, though probably not the direction, of the effects noted by Potts.

Traps seem to be comparatively little used in controlling bird problems in general farming, although Rook traps were traditionally used around poultry. Trapping, however, has proved the most successful way of controlling the rather specialised problem of Bullfinch damage in orchards, which is extensive in some years. This problem perhaps is the classic example of the need for adequate knowledge of a bird's habits before devising control methods. The most significant damage is caused by Bullfinches feeding on fruit buds in winter in years when the seeds normally present in their diet, particularly ash, are scarce. Newton (1972) proposed, therefore, a programme of trapping in the autumn, not in orchards but in woods and hedges close by, to reduce pressure on the species' natural food supplies, so delaying feeding on buds and restricting the damage. Newton found that this approach had two effects: first, it

Fig. 10.1. Number of Woodpigeons shot per km² (dots) and the combined acreage of clover and oilseed rape in the previous spring, 1961–80 (shaded area). From Potts (1981b).

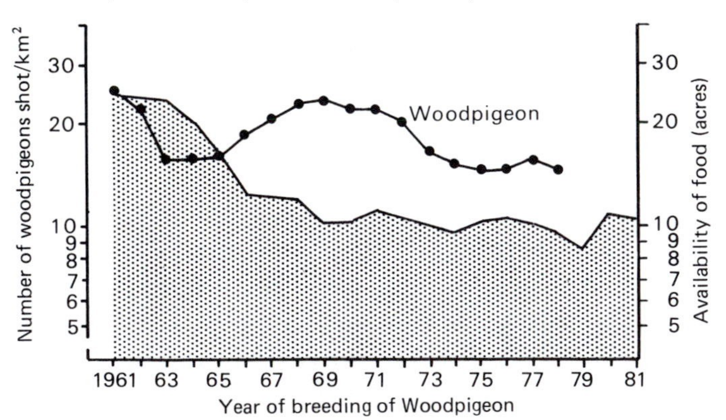

moved the Bullfinch's main period of mortality from winter to the autumn; and second, a very high percentage of the birds trapped were first-year birds, which move about more than adults. The method has proved very successful in controlling damage and is now widely used. It also provides an elegant example of biological pest control which has given good damage control while apparently not affecting the numbers of Bullfinches. It demonstrates clearly the sort of approach needed with all such problems.

In sum, three basic points emerge from this brief discussion of crop protection. It is necessary to study the behaviour of the species concerned since, without accurate knowledge, effective control is rarely possible. Secondly, the aim should be to control or limit damage, not the bird, since the latter is rarely feasible. Thirdly, in the case of numerous and gregarious birds such as geese, alternative food sources are necessary to keep them off crops, a point which farmers, particularly in today's social climate, will probably have to consider with increasing frequency.

Shooting for sport

Shooting for sport is an increasingly valuable asset on many farms and estates. This is a subject about which many people concerned with conservation have understandably mixed feelings. But there is little reason to doubt that shooting has been and remains generally beneficial to the farmland scene. The game concerned are primarily birds, and management which benefits gamebirds generally also benefits other birds. Furthermore, in today's economic climate, one can question whether habitat features and management retained because of their value to game preserving would remain if shooting were discontinued (Barber, 1970). The classic example of this concerns heather and grouse in Wales. Here the increasing emphasis in the uplands on forestry and sheep farming, rather than on grouse shooting, has steadily reduced the area of heather moorland. Yet this moorland is the preferred breeding habitat for some of the most attractive upland birds, such as Golden Plover and Merlin. Other uplands have been similarly influenced (Fuller, 1982). Concern for habitat maintenance in the interests of game has therefore been a force alleviating the worst of habitat destruction on at least some farms. Interest in game shooting provides reason to retain agriculturally less productive habitats on the farm. According to Potts (1970), the rate of hedgerow removal on farms owned by Game Conservancy members was exactly half the national average.

A successful shoot based on wild game requires three things: food and water for game, shelter and cover for game to breed, and a varied terrain for the presentation of sporting shots. Where gamebirds are reared in pens for release, nesting cover may be less important. Most game biologists would add a fourth requirement, that of predator control, but few conservationists as yet regard the evidence for this as convincing (but see below). Food is normally provided in two ways: at feeding stations in woods, hedges and rides, which are supplied at regular intervals, or by planting strips or patches of food plants such as maize or sunflowers. Variants of these themes are to plant strips of crops such as kale and supply food in them, usually scattered in straw, or to tip cleanings from grain-driers in clumps of cover or simply to leave some stubbles through the winter. As with the food sources provided by stock-yards, other birds benefit from these activities and one trailer load of cleanings will keep a wide variety of species happy for months.

Water may sometimes be as necessary as food, particularly in drought or long spells of frost. Most well-managed shoots ensure a regular and unfailing supply for game and this, as with the food supplies, becomes freely available to all birds. Such resources must add to the general interest of any area.

Shelter may be provided simply by keeping and maintaining what is there or by planting anew. The first may assume increasing importance in the future unless hedges slowly die out through lack of maintenance other than an annual trim. The provision of patches and belts of shrub cover is now a widespread practice on farms and estates where the sporting interest is valued and has been encouraged by such bodies as FWAG and the Game Conservancy. Everything benefits, provided the choice of species for planting is sound. Another form of cover is the planting of strips of ground cover for nesting. Similarly, small broadleaved woodlands – often valuable to birds (Chapter 4) – may yield a combined income from hardwood thinnings and Pheasant shoots which may exceed that obtained by conversion to a softwood plantation less favourable to birds. Coppice, for example, is especially valuable to game. Thus many aspects of managing a shooting interest well are an integral part of farm or estate management. As such, it forms one of the best types of conservation management on the farm, since it becomes part of the routine management of the farm.

Predator control and abuse of pesticides

Much of our understanding of the dynamics of Grey Partridge populations in relation to predation and habitat changes is due to the work of Potts (1980, 1984b). More than 200 man-years have been invested in partridge studies so that it has been possible to create a population model that correctly mimics the observed population decline (Fig. 10.2). Since the factors built into the model – predation rates, availability of nesting cover, the abundance of chick food, and so on – result in a good description of the real population of Grey Partridges, the model can confidently be used to investigate what would have happened to Grey Partridges living in a different world – one without any predators, for example.

The partridge model, at its simplest, starts in March with a population of paired birds whose nests suffer density-dependent predation. Density-dependent predation means that proportionally more predation occurs the greater the number of pairs that nest in a given area. The resulting chicks then survive or starve in numbers set by the availability of suitable invertebrate food. The surviving chicks and adults enter the September population and are hunted. Their casualties at this time are again density dependent. Those that survive the winter emigrate prior to nesting and the new breeding population then repeats the whole sequence through the following year.

The population level is most sensitive to losses of females on the nest, particularly due to predation by Carrion Crows and other egg predators which have alternative prey but which still take partridges even though at very low levels. Such predation is estimated by the Game Conservancy model to account for about half of the observed post-war reduction in partridge numbers. Here changes in the intensity of keepering have been important. At densities of 20 pairs/km², for example, predation was 30 per cent less in keepered than in unkeepered areas. Habitat loss, particularly of hedgerow, significantly influences the amount of nest predation. If

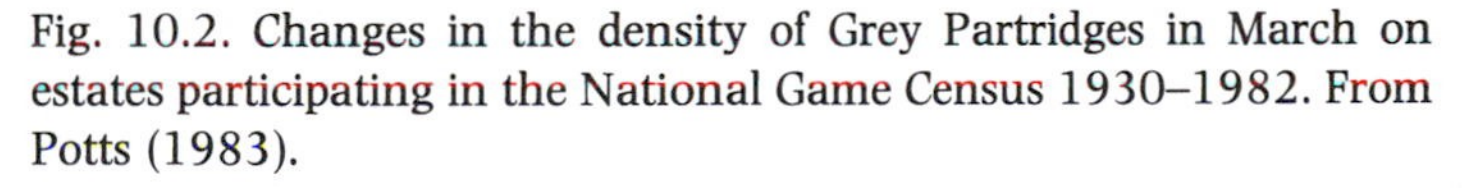

Fig. 10.2. Changes in the density of Grey Partridges in March on estates participating in the National Game Census 1930–1982. From Potts (1983).

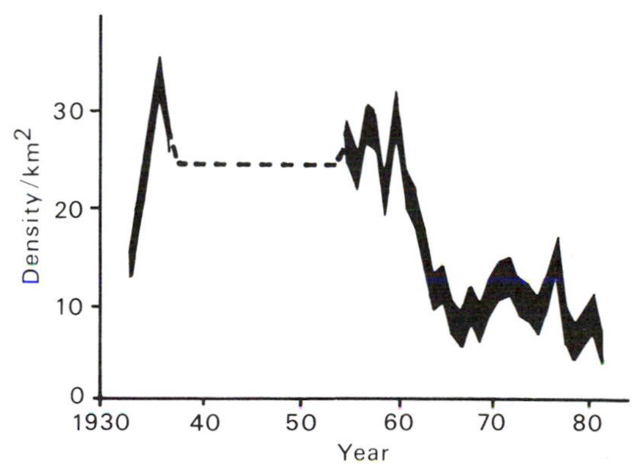

nesting birds are crowded together into small patches of suitable hedgerow, they suffer greater mortality (and the population decreases more) than if they can space out. This effect was confirmed by examining predation by stoats on nests in a Norfolk study site, where it was found that nests predated were indeed spatially clumped. The maintenance of an adequate volume of suitable habitat is therefore important in addition to predator (particularly fox) control by keepers. When the model was used to calculate the density of Grey Partridges that would prevail in the absence of both shooting and predator control, it turned out that up to 30 pairs/km^2 would be sustained if optimal levels of hedgerows were restored.

These findings indicate that habitat loss and predation interact in driving gamebird populations downwards. In the case of the Grey Partridge, the chicks depend on insects and the abundance of these insects – determined in part by the level of pesticide use on the farms – determines the survival of chicks (p. 198). Potts estimated what would have happened to the Grey Partridge population had the chick food supply not been killed off by the insecticidal side-effects of herbicides and found that this factor alone was responsible for as much as 43 per cent of the observed decrease. Red-legged Partridges do not have this strong dependency on insect numbers, however, yet they too declined in the Sussex study area. Red-legged Partridges are even more subject to nest predation, due to their habit of double clutching (Green, 1981). A female will usually lay eggs in a first nest and leave the eggs unincubated whilst, still accompanied by the male, laying a second clutch which she will incubate, the male at that stage returning to incubate the first clutch. The first clutch is left exposed for a period of about 20 days and is therefore at greater risk of predation.

One reason for this double-clutch strategy of Red-legged Partridges appears to be very high losses to predators (corvids, hedgehogs, stoats and rats) throughout the nest stage. In one Norfolk study daily nest loss was as much as three times that of Grey Partridge nests (Green, 1981) and is only partly due to the first clutches being at risk during the laying of the second clutch. The habit among Red-legged Partridges of leaving their eggs uncovered during laying, whereas the Grey Partridge covers the eggs in leaves, may be responsible for these high predation rates. The net effect, however, is that Red-legged Partridges produce on average only 60 per cent of the young fledged by Grey Partridges, despite the production of two clutches by some pairs. Red-legged Partridges would therefore gain more from the control of nest predators by gamekeepers than would Grey Partridges.

The partridge model thus places great emphasis on the importance of predation in reducing gamebird populations trying to survive in an agricultural environment which is deteriorating in food or habitat quality. Several of the problems identified can be mitigated by suitable management on estates where shooting is paramount. Yet the genuine benefits that wildlife can and does derive from field sports, mainly shooting, is often vitiated by the malpractice of some gamekeepers. The destruction of Britain's birds of prey by the game-preserving interest in the nineteenth century is a matter of record (see, for example, Newton, 1979), as is also that it continues covertly today (e.g. Cadbury, 1980b). At the end of the nineteenth century game management was seen simply as a matter of excluding poachers and eliminating all possible game predators (Tapper, 1982). Habitat management was hardly necessary, for hedged fields and crop rotations favoured partridges and Pheasants. Some idea of the historical efficiency of gamekeepers in controlling the numbers of predatory birds can be obtained from Fig. 10.3 which shows the changes in range of three species widespread through Britain in 1800 and heavily restricted by 1900–15, though some range expansion has occurred since then. These

Fig. 10.3. The distribution of the Red Kite, Buzzard, and Hen Harrier in Britain before and after the advent of predator control for game management. Dotted areas are the distributions in 1800 and hatched areas are the distribution between 1900 and 1915. Range extensions since then are enclosed by broken lines and indicated by arrows. From Tapper (1982).

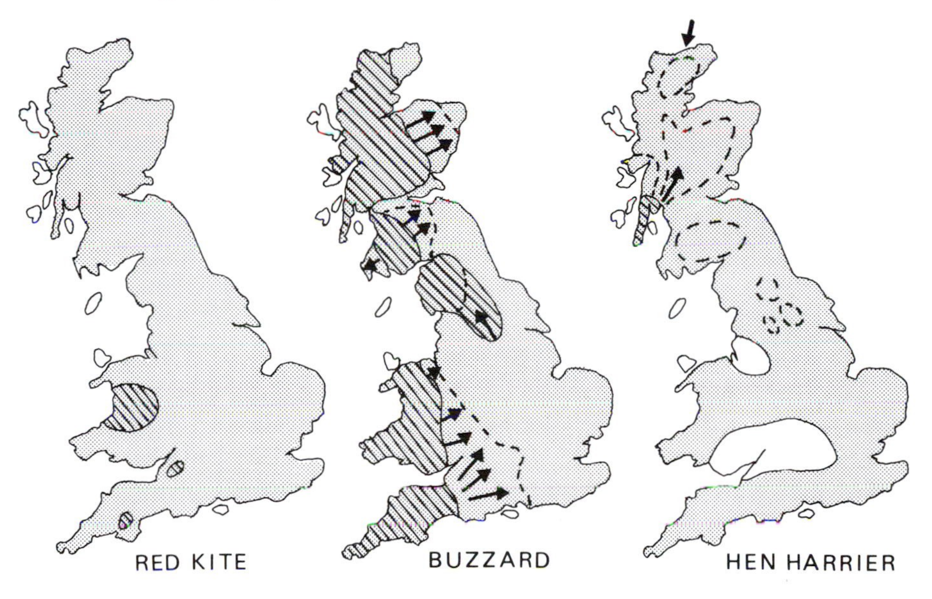

latter-day expansions are due in part to legislation and in part to the great reduction in the number of full-time gamekeepers in the country. They are now too few to control predator numbers with anything like the same efficiency as in the past, even were the methods available to them then still legal. At least 25 000 gamekeepers were at work in 1911 but only *c*. 10 000 in 1920; numbers increased slowly until 1940, when they fell once more, and they have subsequently varied a little around 5000 (Potts, 1980). More important still has been a fall in the amount of farmland covered by gamekeepers. This has fallen from about 50 per cent at the turn of the century to probably less than 8 per cent now.

Over the same period, however, many small predators, both mammalian and avian, have increased in numbers (Fig. 8.19, 8.20). One response has been a growing abuse of farm insecticides to destroy predators both in game-preserving areas and in upland sheep-farming areas. This subject has been fully discussed by Cadbury (1980b). He found that 52 per cent of the incidents reported of the misuse of poisons in the countryside between 1966 and 1978 involved agricultural insecticides, nearly all of them organophosphorus compounds, and 45 per cent involved one chemical compound, mevinphos. In 1981 the figures were 44 per cent of 128 reported incidents involving agricultural insecticides, 41 per cent of them mevinphos. The legitimate use of mevinphos is restricted to orchard and vegetable crops and Sly (1981) estimated that it was thus used on only 2332 ha of crops in England and Wales. The total amount of active chemical ingredient is 0.52 tonnes and the cost per hectare given in the Crop Chemical Guide 1982 is £7.40, indicating a legitimate market of well under £20 000. Regional analysis of the records by Cadbury indicated that mevinphos was being abused in areas lacking legitimate commercial demand, for the seasonal occurrence of incidents suggested a strong bias towards misuse in game preserving and sheep farming, activities in which agricultural insecticides of this type have no place. Indeed, within the game stronghold of Norfolk relatively few incidents of misuse were recorded, presumably due to more professional attitudes by full-time gamekeepers there. Otherwise the distribution and recurrent pattern of misuse suggest that such abuse is becoming an increasingly regular management tool.

The most serious aspect of the problem is the indiscriminate way that such chemicals are abused. Thus many working dogs are killed by baits apparently intended for foxes. The improper use of grain baits has also killed many non-pest bird species. Cadbury stresses the massive doses being laid in baits and notes that, of 53 bird species involved in the incidents reported, only 14 were listed on the Second Schedule of the Protection of

Birds Act and, of these, only two or three can, in fact, fairly be described as pests of crops. Raptors have been the main victims. Although the impact of these activities on raptor populations has locally been quite marked, particularly for larger species such as the Buzzard (Picozzi & Weir, 1976), for conservation the real danger in this abuse of farm chemicals lies in increasing existing prejudice against field sports. We can ill afford to lose the real advantages that properly conducted field sports offer the conservation of wildlife in the farmed countryside generally.

In Scotland deaths of wildlife due to agricultural chemicals are monitored by a Wildlife Incident Investigation Service (WIIS) set up in 1972 by the Department of Agriculture and Fisheries for Scotland. This followed an incident in which over 500 Greylag Geese and Pink-footed Geese were poisoned by the organophosphorus compound carbophenothion (Bailey *et al.*, 1972). Hamilton *et al.* (1981) provide a review of the misuse of agricultural chemicals in Scotland, as investigated by the WIIS over the period 1973–79. The principal victims of these kills over the period were gulls of various species (193 birds in all, including 55 identified Common Gulls and 120 *Larus* spp.), Buzzards (34), Jackdaws (32), as well as 10 or more of Rook, Crow, Woodpigeon and Pheasant. Three poisons – mevinphos, alphachloralose and strychnine – accounted for 94 per cent of all incidents, with two-thirds of them occurring in the period March–June. These figures apply only to deliberate misuse of poisons, for when the WIIS started in 1972 large numbers of mass bird casualties were still occurring in the principal wheat areas. This was due largely to the use of dieldrin or chlorfenvinphos as seed treatments of winter wheat as a preventative of wheat-bulb fly. The concentration of incidents into the spring probably reflects a desire to protect newly planted crops and young game or livestock. Aldrin and dieldrin were withdrawn as seed treatments in 1975 following further wildlife casualties.

These problems underline the difficulties of controlling the use of poisonous substances in the countryside, particularly by a voluntary scheme such as the Pesticides Safety Precaution Scheme. Such a voluntary system demands a standard of integrity among all concerned which is evidently not always available and proposals for more stringent sanctions and controls, which will have the force of law, are before Parliament even as we write.

Summary

Although many species eat crops, few behave as pests. The principal species that do so are geese, pigeons, corvids and Starlings, with others such as

Skylark and Bullfinch troubling certain crops. Modern agriculture tends to promote pest behaviour by limiting natural food supplies but there are important natural restrictions, e.g. seasonal limitations to crop susceptibility, traditional patterns of behaviour, etc, to most pest problems. An important exception is the stealing of animal food from indoor stock systems. The most successful preventive measures are efficient scaring over crops and restricting access to stock feed. Population control by shooting may have been more successful in controlling Woodpigeon numbers than previously realised. Gamebirds are a valuable asset on farms and estates and the management of shoots makes a worthwhile contribution to the conservation of wildlife. However, game populations, particularly of partridges, have been severely affected by modern farm management and by changes in predation patterns. Predation has increased with a decline in the number of gamekeepers and the abuse of pesticides has increased in parallel.

11: The outlook for farming and birds

Much of the material presented in this book has shown that farmland bird communities in Britain are influenced both by habitat and by food supplies, though the relative importance of each may alter seasonally. One may therefore be inclined to assess the outlook for these bird populations as a function of habitat and food. In our view bird conservation in agriculture would be perpetually out of date with such an approach, constantly responding to the specifics of the latest changes in agriculture without strategic signposts for the future. Our analyses show that rapid nationwide changes in the pattern of agriculture in Britain have had major impacts on the bird populations of whole regions. Hence bird populations nationally will be more influenced by changes in agricultural policy and in technology than by the sum of conservation-oriented measures on individual farms. Here we outline some implications of such dominance by policy and technology.

It seems clear that the political and economic background in which farming operates will change sharply in the next few years, as the EEC grapples with the problems of its Common Agricultural Policy (CAP).

Common policy will probably be increasingly difficult to devise as the EEC grows to embrace countries as diverse in climate, geology, agricultural products, and economic and political structure as, for example, Greece and Britain. Furthermore, the present level of monetary support for agriculture looks increasingly anachronistic, at least for the major countries within the EEC. In theory such systems of support provide the machinery to set production fairly closely in line with needs but in practice they rarely work accurately for long. Too many factors, particularly climatic, govern the level of production but are outside the control of the administrator and farmer. In addition, systems of capital grant aid weaken or destroy the link between investment and return on money invested and favour structural changes that could not be undertaken if financed purely out of revenue earned. This automatically encourages overproduction. The CAP also differs sharply from the 1947 Agriculture Act in having dual objectives: of promoting food production and of maintaining rural communities and their standards of living. The 1947 Act concerned itself only with food production and was therefore partially sensitive to adjustments. The extent to which the CAP is now bedevilled by overproduction and the difficulties of curbing it may be a measure of the incompatibility of its two objectives.

The farming systems and methods of the nineteenth and early twentieth centuries not only created much of the landscape we now wish to preserve, the routine farm management involved incidentally maintained it. In particular modern farming no longer does the latter. Instead it is evolving new habitats of much simpler form and poorer in species. This results from the basic change in farming from generally mixed systems and methods to specialised ones based on the application of technology. In the preceding chapters we showed that the specialisation occurs in four dimensions: it occurs in the field with the wide-scale elimination of organisms competing with or living in or feeding on crops; it occurs on the farm with the elimination of labour-intensive operations and the concentration of capital on fewer enterprises; it occurs regionally with the concentration of crops into the most favourable area for them; and it occurs more widely geographically with the increasingly clearcut division of an arable east and a pastoral west. Modern communications and teaching also ensure a very rapid and uniform application of technology. Such specialisation is almost certainly the single most important factor affecting the successful conservation of wildlife in farmland. It results in both a steady decline in the diversity of ordinary farms and a sharp decline in the farm's tolerance of unmodified habitats, such as wetland, because specialised farming can only make decreasing use, if any, of such areas.

Table 11.1 summarises the decline of mixed farming in England and Wales between 1963 and 1977, when MAFF gathered statistics on farm types. The categories used were fairly broad but they nevertheless serve to demonstrate the growth of the specialist farm in two important areas. The loss of general mixed farms is also striking. Because of the gearing effect of an overall decline of farm units, the change in this category in the period between 1963 and 1975 represents a fall of *c.* 72 per cent and such farms were then disappearing at an average rate of 1300–1400 annually. Yet one of the recurring themes in any detailed study of farming and birds is the value of mixed farming to farmland birds. The distribution of wintering birds illustrated on p. 35, the way in which birds exploit farming operations for food, the way that they exploit the presence of stock (particularly cattle), and the importance to them of several seasons of cultivation are all examples. Without this variety in cropping and management the preservation of isolated areas of semi-natural habitat seems most unlikely to provide a satisfactory basis for farmland bird populations, however suitable such a preservation policy may be for other wildlife. This is a view also expressed by Mellanby (1981).

Farmland bird communities tend to consist of two parts, a group of common and widespread species which form the basic farmland bird community, and a variable number of scarcer birds which are attracted into farmland by the presence of areas of more specialised habitat. While populations of the latter have changed very much in line with changes in the stock of semi-natural habitat – old grassland, hedgerows, woodland, and so on – within farms, the main group shows an interesting pattern. Many of these species recovered steadily from the impact of the 1962–63 winter which coincided with the start of the CBC, reached a peak in the early 1970s, and have declined sharply since. In terms of area affected, habitat loss was most marked in the 1950s and 1960s but many of the management changes we have discussed in earlier chapters have developed in the period from the late 1960s to the present. The general decline of common farmland species has thus coincided with the period of changing agricultural management rather than with the period of habitat loss, an argument supported by many of the detailed species studies discussed earlier.

Although management changes have been behind the major changes recorded of farmland bird communities in recent years, for many conservation bodies national agricultural policy is the principal factor which determines the nature and progress of farming and its impact on the countryside. Many of the points discussed in this book suggest that this

Table 11.1. *Changes in the types of farm holdings in England and Wales, 1963–77*

	Specialist dairy as % of cattle farms[a]	Specialist cereals as % of arable farms[b]	% of total cereals grown on cereal farms	General mixed farms as a % of all farms[c]
1963	36.3	—[d]	18.8	6.7
1964	37.1	25.3	18.8	
1965	37.2	29.5	21.9	
1966	40.1	33.2	24.7	4.9
1975	46.1	39.8	27.1	3.1
1976[e]	48.3	29.0	19.0	
1977[e]	49.8	29.8	20.0	2.7

[a] Cattle farms include (a) specialist dairy farms (over 75% in dairying); (b) mainly dairy farms (50–75% in dairying); (c) livestock rearing farms, mainly cattle (over 50% in livestock of which over 75% in cattle); and (d) general livestock farms (over 50% in rearing cattle and sheep).
[b] Specialist cereal farms are those in which over 50% is in arable cropping of which over 50% is in cereals.
[c] General mixed farms are those with less than 50 per cent in any one enterprise.
[d] Not recorded.
[e] This survey was redefined in 1976 and ceased in 1977 but the trends shown continue.

view needs modifying. Agricultural policy probably has most influence on the status and distribution of semi-natural habitats on farmland but it is technology which largely governs management. Management is, after all, determined by technical considerations such as the incidence of disease or the need to control weeds, matters which are always present and quite independent of policy decisions. It is important to decide how far these two factors of technology and policy act separately. Clearly they are linked in the sense that the development of farm technology has been hastened by the policy decision to encourage farming. Policy is also influenced by technology as price levels anticipate technical capacity and therefore influence the rate at which that capacity is applied. Furthermore, a policy of sustaining a rather narrow range of farm products must have had a negative effect on technology by limiting the incentive to explore new crops and crop uses, which may be increasingly important to farming in the future. Nevertheless, technical innovation in farming continues all the time. For example, the first applications of the internal combustion engine and of modern chemical technology to farming date from the depression years before 1939. It can be argued that without such technical innovation changes in agricultural policy since 1947 would have had far less ecological impact, and conversely, that the technical innovations of the past 25 years would have been attractive to farmers anyway, as a way of improving profitability. Thus the two factors, policy and technology, are linked by circumstances rather than of necessity.

One of the fundamental questions for the future of farmland birds, therefore, is whether this shift in emphasis from major habitat losses to major management changes as the main cause of impact on farmland birds will continue. Three reasons suggest that it will. First, there is a declining stock of semi-natural habitat and much of what remains will be increasingly vigorously defended. Secondly, changes in the economic circumstances of farming will very probably accelerate the application of technology. Thirdly, the only sensible course available to farmers when traditional markets are shrinking and when Government commitment to farming is being reduced is to broaden their economic base.

At present the changes in policy that appear to be set are the imposition of quotas on milk production; a decline in the value of cereal support prices which, together with the present level of overproduction, has greatly depressed prices; and a major reduction in the grants available for farm improvement. It seems fairly clear that farming faces a period of steady decline in the return from its traditional operations. Indeed, a report in *The Observer* for 9 June 1985 suggests that net agricultural income in Britain has fallen from £567 million in 1970 to £316 million in 1983, at constant (1970) prices. Unlike most commercial enterprises, farming has little control over its main markets. Farmers' main management resource in an era of declining returns, therefore, is to cut the cost of production and to increase yields.

The problem with such a prescription, however, is that it is eventually self-defeating. It is largely because farmers generally have reacted to economic pressure or stimulus by increasing technical inputs and raising yields that the present surplus situation exists (Fig. 11.1). This has provided the major impetus to evaluate new crops, resurrect discarded ones and develop new uses for existing ones. Pulses for animal feeds, durum wheat for pasta, linseed for industry, lupins for oilseed, and evening primrose for the pharmaceutical industry are all crops at present being developed, and a much wider range of plants is the subject of experiment for food, forage, industrial, and medical applications. Energy provides the greatest opportunity in the widening of crop use. All farms producing arable crops have opportunities to make major savings in energy costs by using crop byproducts such as cereal and rape straw for heating. Other experimental crops are grown purely as fuel crops, while rape and other vegetable oils have major potential as engine fuel. Stock farmers too can obtain similar advantages from gas fuels derived from animal waste, a technology which is well advanced both in Europe and in North America. Nor should we underestimate the speed at which such new crops and crop uses will be developed, when oilseed rape, the first such crop to be introduced, has mushroomed to the rank of the fourth most widely grown arable crop in Britain in a decade.

Farming now represents a large public investment and, as a primary industry, is a major customer of others such as the farm machinery industry. Thus it seems probable that the EEC will promote the development of alternative crops and at the same time restrict traditional ones. Such a policy should have the effect of greatly improving farm self-sufficiency and the diversity of cropping. The policy is also politically attractive, for it offers a positive investment approach and the best chance of dealing with the economic problem of the cost of overproduction. The policy is attractive, too, because broader crop use and byproduct use particularly demand capital investment in new kinds of plant and machines on the farm, thus benefiting machinery manufacturers who are already more severely affected by an incipient depression in farming than most farmers.

The major drawback of the policy is that it only works where land and climate allow cropping. This factor makes it very likely that the present

Fig. 11.1. The effects of high technical inputs on yields: (a) The agrochemicals used on cereals at Oakhurst Farm in the 1980s; (b) Yield in wheat grown under pre-chemical (left) and post-chemical (right) conditions.

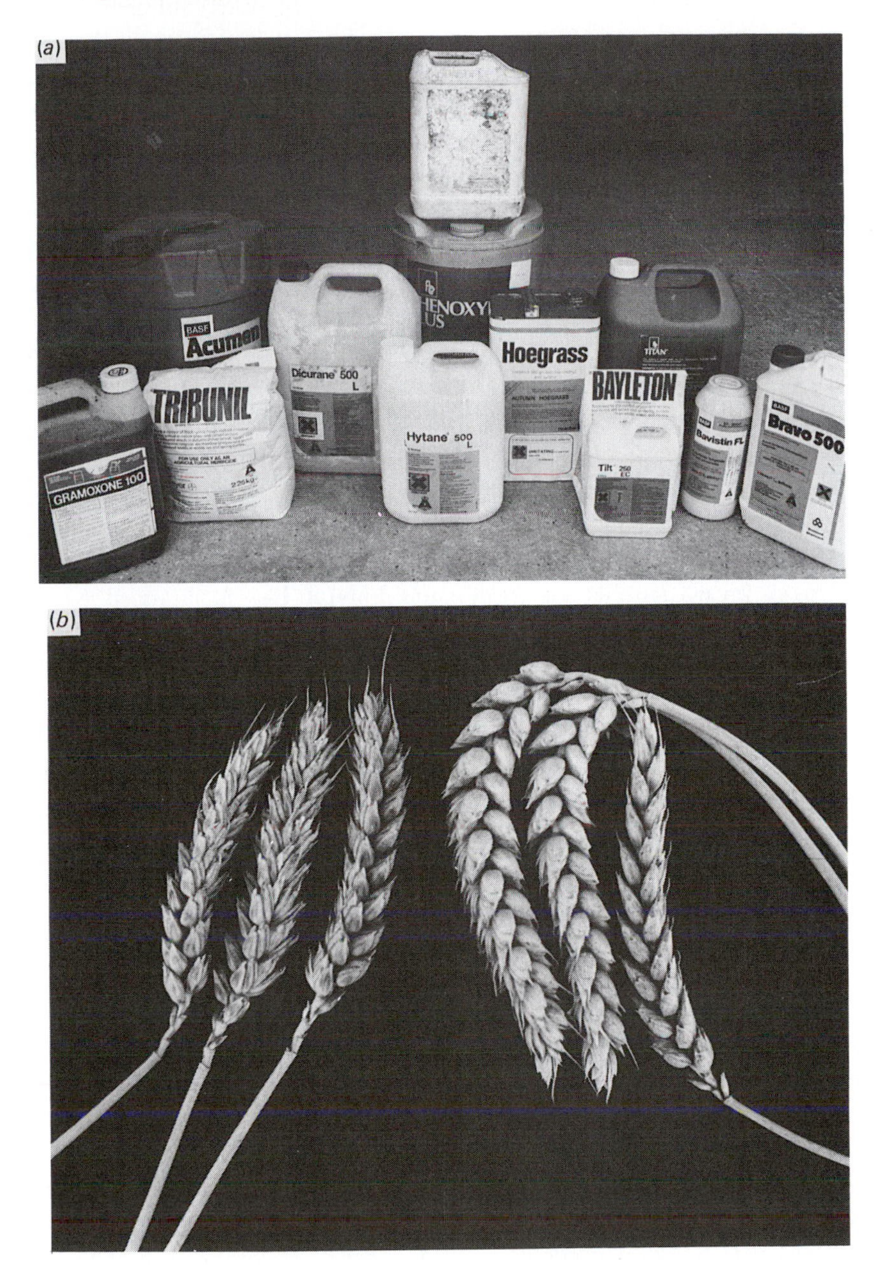

pattern towards concentrating cropping enterprises in eastern Britain and grassland farming on the uplands and in the south-west will continue steadily. Indeed this pattern ought to be encouraged as one way of alleviating the problem arising from the increasing disadvantages faced by upland farming as agricultural support is reduced. It could meet the basic aims of the alternative strategies proposed for the uplands by conservation bodies, which envisage maintaining traditional agriculture there (e.g. RSPB, 1984). Without some provision these areas will become dangerously exposed and may well see a major drift out of farming altogether. Blanket forestry would be the most likely land use to fill the vacuum thus created.

The June Census statistics show that the rate of gross habitat loss deriving from the modern development of farming has slowed markedly in recent years. Thus permanent grass and rough grazing in England and Wales declined by an average of 88000 ha annually between 1932 and 1963, by 46000 ha per annum between 1964 and 1973 but by only 12000 annually between 1974 and 1980. The loss of hedgerow shows a similar pattern, for the earlier rapid loss has been offset by new plantings. Both changes reflect, of course, the steady decline in the area available for reclamation. Recent concern has focused on the wisdom of suffering further loss of interesting habitat to farming when many farm products are in severe surplus. In effect such habitat loss is financed by the State. This concern expresses itself through local resistance to major farmland improvement schemes and through measures such as the Wildlife and Countryside Act 1981 which has introduced the principle of paying Government grants or compensation to have those areas valuable to conservation managed for their scientific interest, albeit with mixed success to date.

The process has also been influenced by the creation of a countrywide network of Farming and Wildlife Advisory Groups (FWAGs). These groups have had much success in encouraging farmers to plant, make, or maintain small areas of interesting habitats on their farms, particularly ponds and copses, the latter often to replace hedges. Comparatively insignificant in themselves, such areas have little impact on any farm's economy but, in total, they provide a very useful addition to the stock of semi-natural habitat in farmland. A major disadvantage of such areas, however, is their very fragmented nature. We note (p. 119) the importance of linking such plantings through a properly managed hedge system to circumvent the results of fragmentation and would stress again the necessity of looking at the layout and management of such a lattice in the face of the working needs of modern farms. The probability of further major technical innovation makes this doubly urgent. Game management offers other possibilities, for shoots have long grown small areas of crops specifically for game and these provide opportunities for other birds (p. 228). No one as yet appears to have explored the idea of expanding this theme for more general conservation, except in the case of refuges to draw geese off economic crops (p. 221). Nor have the needs of wintering and passage birds received much consideration.

It is probable that the Government regards the creation of a countrywide network of FWAGs as the best option for promoting conservation in farmland. Their greatest value, however, lies in encouraging sympathetic attitudes and gaining time, but whether these attitudes will actually survive policies which in effect reduce farming prosperity is uncertain.

In sum these ideas suggest that farming development over the next few years will include three features – continued polarisation with an increasing concentration of grassland in the west and arable in the east; an increasingly intensive application of technology to contain production costs; and a wider range of cropping. If this reading of the future development of farming is correct, then the greatest ecological changes deriving from farming will evolve far more from management changes than from habitat loss. Changes in rotations and in crops grown, further changes in the balance of grassland and tillage, changes in the timing of cultivation and in cultivation methods (perhaps particularly the scale of herbicide use), changes in the methods and timing of harvesting of crops, the increasing use of fungicides and insecticides (particularly in cereals), and the way in which stock is kept have been identified in this book as major factors affecting the numbers, distribution and success of farmland bird populations.

It is the common species – gamebirds and other groundnesting species, ground-feeding species, the common farmland raptors, buntings, and many finches – that will be most affected (not always deleteriously) by these changes. We would recommend, therefore, that the main emphasis in conservation in farmland should in future be focused on these problems quite as much as on the problem of habitat loss. Here it is important to remember that farmers are interested in farming, i.e. the business of raising crops and stock to the best of their ability, and not in managing land for other purposes. Even those who have a strong interest in wildlife conservation and partly manage their farms to that end nevertheless remain primarily farmers. This is one of the main obstacles to the successful implementation of many of the alternative strategies canvassed by conservation bodies. Furthermore, it is as well to remember that farming has a very long tradition of striving to improve the productivity of land in the only way it understands (raising crop yields) and of this being regarded by all as an entirely admirable goal. It is essentially difficult for farmers to grasp modern attitudes towards this or to move towards the ecologist's concept of productivity. In the view of conservation bodies, amenity and conservation are the main alternatives to food production in land use but the increasing spread of alternative crops and crop uses being studied and developed makes it certain that this is not so; we really must not underestimate the speed with which new technology will be applied. This, together with our poor knowledge of the impact of the modern day-to-day operations on birds, may be the most important feature influencing future conservation attitudes to farming and birds.

Summary

Present trends suggest that technology may now be a greater influence on farming than is national agricultural policy. Management has become more important than habitat loss in terms of impact on bird populations. Future farm policy seems likely to aim at developing new crops using novel technology and management whilst restricting the overproduction of existing crops. This, together with the modern concern over habitat loss on farmland, may slow the further reclamation of remaining habitat but will introduce new regimes affecting birds resident on agricultural land.

Appendix 1: The Common Birds Census

The Common Birds Census is an annual census of bird population levels on sample census plots on farmland and in woodland in Britain and is organised by the British Trust for Ornithology (BTO). The farmland CBC began in 1962 as a method of monitoring populations possibly affected by toxic chemicals. The census procedure involves volunteers surveying their census plots between eight and 12 times each breeding season. On each visit all bird registrations providing evidence of breeding behaviour of the species present are mapped on a large-scale map, one per visit. The data for each species are subsequently transferred to separate species maps. The registrations on these species maps tend to fall into more or less discrete clusters corresponding to territories of individual pairs and a count of these provides a census estimate. A degree of subjectivity in the assessment of these clusters is minimised by the use of trained BTO staff analysts for this task. In principle these counts can be converted to density estimates and averaged over all plots censused each year, to provide an annual assessment of population level. In practice, a number of difficulties require that the data be used as an index of population level rather than as an absolute census.

A number of colonial or semi-colonial species are not amenable to censusing by the mapping method of the CBC and most observers do not attempt to census them. Among the common farmland species affected by this are Rook, Woodpigeon, Starling and House Sparrow. Where it is practical to do so, as with Starlings, some observers report nest counts of such species and where this approach is maintained consistently these results can be used in the computation of the CBC index. Otherwise, these species are underrepresented in the CBC, though their presence as breeding birds is normally noted.

The CBC index is based on the pairing of census results from each plot in successive years. This procedure eliminates difficulties due to two problems: census accuracy and observer turnover. Census accuracy varies between species for a number of reasons – some species are more conspicuous than others, some populations contain transient birds present on some but not all censuses, and the ability of the census worker to detect birds may differ between workers. The number of clusters established for each census plot is therefore only an approximation of the population actually present. The absolute census efficiency of each census worker is unknown, and would need to be calibrated individually and for each species present on that plot if the census results were to be corrected to true densities. This has been impractical. The effects of observer turnover are also related to this uncertainty of absolute census efficiency. Each year some census workers leave the CBC scheme and are replaced by other volunteers censusing similar plots elsewhere. As a result, the average census efficiency may vary from year to year. However, by pairing the results from the same observer in successive years, an estimate of the relative change in population level on that plot can be obtained. These estimates are independent of observer census efficiency, provided that this efficiency does not change between years. This is a much less restrictive requirement than needed for absolute census efficiency and this requirement is adequately met by participants in the CBC scheme (review in O'Connor & Marchant, 1979). As a result, relative changes in population level are known with greater precision than are absolute population densities. Population trends are therefore reported in the form of a CBC index based on arbitrarily setting the 1966 index to 100. The indices for other years are then proportional to population densities, at least to a first approximation.

In this book we adopted the same procedure to create population indices for regional and other subsets of the CBC data. The CBC scheme normally has about 100 census plots available for each year's index and this intensity of sampling is more than adequate to avoid any significant problem of chance accumulation of errors in the index. For regional and other subsets sample sizes were smaller and this risk was greater. We therefore checked all index results reported here against trends in the less precise annual estimates of cluster densities for the subset concerned. These estimates are less precise, so that each data point has a larger uncertainty than has the CBC index for the year, but are less prone to accumulation of error. We originally intended to standardise on 1962 as the reference year (index = 100) for these subset indices but we experienced the problem of some species being slow to recover from the effects of the severe winter of 1962–63. In some early years, therefore, territories present within some samples of CBC plots were too few to permit calculation of an index for that year. In these cases we used the next available year as reference year instead.

As already noted, absolute densities are less precisely established by the

CBC method than are year to year changes. Nevertheless, O'Connor & Marchant (1979) provide crude estimates of the uncertainty of estimation of absolute densities for various species and show that about one in four territories (irrespective of species) might be missed through less than perfect census efficiency, though the rate varied markedly between species. Census errors of this amount are not large in comparison to the two–fold to six-fold variation in population levels observed in different species over the 23 years of the CBC. We have therefore used the density estimates obtained from the CBC for some purposes here, mainly in Chapter 2. Here, the pooling of density estimates across all censuses available between 1962 and 1980, e.g. in Fig. 2.1, is at first sight objectionable, since some census plots were surveyed over many years whilst others were surveyed only over two years. However, the alternative analyses possible were as problematical, for the year to year variation in population levels precludes the use of any one year as a suitable single sample, whilst the use of several single-year analyses spread through the period would retain the problem of some sites being common to some or all of them. In addition, Fuller, Marchant & Morgan (1985) have shown that, over the period of the CBC, agricultural practice and land use on CBC plots within lowland England have been approximately in line with the patterns recorded in the MAFF June Censuses, so that what has happened on the CBC plots has been fairly typical of what has happened elsewhere. The use of pooled data in Fig. 2.1 and related analyses therefore provides a simple, even if approximate, picture of the general features of farmland bird communities over the period.

We have presented population trends here only if we were satisfied that trends in appropriate control areas differed significantly from those attributed here to the specified agricultural practices. We have chosen not to present the results from the control areas themselves because of the sheer complexity of defining and justifying such areas. The difficulty is that, whilst a subset of counties homogeneous in agricultural practice could be defined in a straightforward fashion, the counties omitted were then almost invariably agriculturally heterogeneous and at times dominated by counties in which other changes affecting bird populations were taking place. Hence we could not use the pooled results from these areas as a single control without specific discussion of the influence of farming changes other than the one of immediate interest. We therefore adopted a different approach, of reviewing simultaneously the population changes recorded in several subsets, some of them mutually independent but others overlapping in composition. Sample sizes varied between species within these subsets, so that in some cases no analysis was possible, but, subject to this, we have presented data here only if justified by reference to controls appropriate to that species. Copies of the relevant printouts and plots can be made available to qualified researchers.

We examined the impact of agricultural changes on bird population by computing CBC indices within subsets of counties selected on the basis of uniformity of agricultural practice (or change therein). The composition of these subsets is explained in detail in Appendix 3. Counties in which several major changes in agricultural practice have taken place simultaneously, or in which a diversity of farming regimes have been practised simultaneously, were excluded from these samples. Classification was on a county-wide basis since this was the basis of the MAFF statistics, so this meant that within the areas dominated by particular regimes individual CBC plots could be dominated by some minority regime. In theory our approach might conceal differences associated with the dominant regime (for example, by including the results from some bird-rich pastoral farms with those from an otherwise species-poor sample of cereal farms). In practice, however, the differences between the bird communities of the farming regimes reviewed proved robust enough to emerge despite the gross nature of our analysis. In addition, we know from the study of Fuller *et al.* (1985) that farming practice on CBC plots in different regions in lowland England has been broadly representative of the regional pattern of farm management in the country since the inception of the CBC scheme. Hence where in our sampling of agricultural practice we include a proportion of the 'wrong' type of farm, this proportion in fact reflects the mixture of farms in the area. Our results are thus representative of the bird communities in regions dominated by particular agricultural practices.

Appendix 2: Nest records and their analysis

The BTO Nest Records Scheme began in 1939 but only data for the period from 1962 onwards were considered here, to parallel the CBC data. The nest records are contributed by volunteers who submit a nest history for each nest they find. Pre-printed cards are used, on each of which data relating to species, place, altitude, nesting habitat and site are entered, together with counts of eggs and young made on each visit to the nest. Additional information can be supplied as to the state of eggs and young on each visit and on the outcome and cause of failure of the nest, where these are known. These cards are lodged with the BTO at the end of each season.

At the Trust the data are entered into computer files and analysed using specially written software. Automated analysis is on the basis that successive visits provide ever tighter limits to possible events of interest, such as the date of laying, of hatching, and of fledging. In many cases the observer's estimates of chick ages or of whether incubation had started further narrowed the 'window' within which these events occurred. Where the recorded information is adequate, each of these points can be established precisely but otherwise only an upper and a lower limit can be determined. In this book we accepted all dates that could be established to within a five-day period (except 10 days in Fig. 3.4). Clutch size can similarly be established. Other information on the card is also coded, so that nesting habitat and outcome of the nest, for example, are available for use in subsequent analyses. One restriction here is that for most species the geographical location of the nest is presently coded only to county, so that for analyses of breeding success in relation to agricultural practice this was the limit of resolution possible.

The main problem with using nest record cards collected by the BTO membership at large is that many of these cards relate to nests found some way through the nest cycle. These nests are by definition more successful than were those nests that failed before they were found and recorded by a BTO observer; simple estimation of the proportion of nest records that were eventually successful is thus biased in favour of success. Ideally only nests found at the early stages of building should be used but many of the nests on record in the BTO collection were found later in the nest cycle. A method developed by Mayfield (1961, 1975) gets around this problem by computing success rate not as the proportion of nests that are eventually successful but as the proportion of nests surviving per unit time. Since most nest record cards describe two or more visits to the nest we can take a sample of cards, compute the number of days the nests were under observation and count the number of nest failures recorded over that time. This provides an unbiased estimate of the rate of nest failures per day. The method assumes that the chances of a nest failing are constant within the incubation period and are constant within the nestling period (though these two constant rates may themselves differ), an assumption that seems to be valid in practice (Klett & Johnson, 1982). A previous example of the use of this method for British birds can be found in Bibby (1978). Mayfield's method has subsequently been validated and developed by Johnson (1979) and by ourselves.

In the previous major study of the value of farmland as a breeding habitat, Snow & Mayer-Gross (1967) defined breeding success as the percentage of nests that resulted in at least one fledged young. We adopted a different definition of breeding success, however, since such a definition of breeding success gives an optimistic estimate of how good a habitat is for birds. The definition does not allow us to distinguish between habitats in which nestling starvation occurs at different rates. Suppose two habitats each suffered, say, 25 per cent nest predation but that in one the parents could find enough food to rear an average of four fledglings per surviving nest whilst in the other habitat nestling food is so scarce that each pair can raise only a single fledgling. Using the nest-based definition of Snow & Mayer-Gross (1967), the two habitats appear equally good: each achieves 75 per cent success. Yet the former produces four times as many young as does the latter. Where predation is the major cause of nest failure, as in Snow & Mayer-Gross's study, their definition is not misleading in this way but where feeding conditions vary substantially between habitats some other measure of success is needed. As we were particularly concerned in our analyses to detect any possible impact of changes in farming practice on breeding conditions for birds, we followed Ricklefs (1969) in adopting as a more sensitive measure of breeding success the rate of survival of individual nestlings. Details of the computational procedures will be published elsewhere. Finally, it was necessary to extend the same approach to the calculation of egg loss rates. Ricklefs (1969) showed that infertility of eggs such as might be brought about by exposure to pesticides requires a similar treatment of partial loss of nest contents as does nestling starvation. We therefore computed egg loss rates in a manner exactly analogous to our calculation of nestling mortality.

Table A2.1 illustrates the effects of using these different measures of breeding success with the farmland nest records for seven common hedgerow species. Data from the study by Snow & Mayer-Gross (1967) as to the nesting success of these seven species breeding on farmland near Oxford are included for comparison. The effect of the different definitions of success in measuring overall success is quite marked, by a factor of two in the cases of Dunnock, Song Thrush, Whitethroat and Chaffinch. In other words, in these species about half the eggs or young in those nests that survive total destruction nevertheless die for one reason or another. Loss rates during incubation and nestling periods are separately tabulated in Table A2.1 and show that these partial losses occur mainly during the nestling period, though for three species – Dunnock, Song Thrush and Whitethroat – a significant proportion of individual eggs fail to hatch. Even within these species, though, greater proportions of nestlings than eggs are lost, probably mainly to starvation (Ricklefs, 1969; O'Connor, 1984b). The proportion of young thus lost varies from one in every two Whitethroats down to only one in eight Dunnocks.

Although these figures for partial losses are large, they are in fact additional to a substantial degree of total nest failure (Table A2.1). Overall nest success varied from 42 per cent among Blackbirds to 78 per cent in Whitethroats. Total nest failures therefore accounted for between a quarter and three-fifths of all losses, due mainly to predation. The similarity of the nest loss figures between the incubation stage and the nestling stage suggests that much of this predation is due to predators searching out the actual nests rather than locating the nests by tracking adults as they bring food to the young. Moreover, the individual species varied widely in their success in a way apparently linked to the conspicuousness of their nests. Chaffinches, Blackbirds and Song Thrushes, which have bulky or exposed nests have low rates of success whilst the most successful species, the Whitethroat, has a well-concealed nest built only in well-developed cover. Even for this last species, nesting success varies with nest height, so that the lowest and best concealed nests are the most successful (Mason, 1976). Although we attribute these total losses to predation, some such losses are undoubtedly due to weather-related desertion. For example, whilst 52 per cent of Whitethroat nest failures were due to predation, a further 32 per cent were due to desertion triggered either by a predation attempt or by inclement weather (Mason, 1976). Adverse weather is probably underestimated as a source of nest failure since its correlates may be shared with

Table A2.1. *Various estimators of nesting success and its components in some common birds on farmland*

Species	Percentage survival during				Overall success		Nest success estimated from intensive study[e]
	Incubation period		Nestling period				
	Nests	Eggs[a]	Nests	Young[b]	Nests[c]	Eggs[d]	
Dunnock	69	58	55	48	55	28	34
Blackbird	63	56	68	49	43	27	28
Song Thrush	68	58	70	41	47	24	26
Whitethroat	90	71	87	45	78	32	65
Chaffinch	68	59	70	41	47	24	17
Greenfinch	65	58	75	52	49	30	36
Linnet	67	63	64	50	43	31	43

[a] Allowing for eggs lost from partially successful nests.
[b] Allowing for young lost from partially successful nests.
[c] Percentage of nests fledging at least one young.
[d] Percentage of eggs resulting in a fledgling.
[e] From Snow & Mayer-Gross (1967).
Based on analysis of nest record cards for 1962–80 and on Snow & Mayer-Gross (1967).

those of predation. For example, in a study by Osborne & Osborne (1980) Blackbirds' nests in the higher and more exposed positions – features that might expose them to greater predation – were found to fail more frequently in bad weather than did nests lacking these features.

Table A2.1 thus brings out the importance of adequate definition of breeding success in environments where feeding conditions are a significant source of mortality. In all cases the inclusion of partial losses of eggs and of nestlings yielded a less optimistic estimate than did the measure of nest loss used in the earlier studies and additionally provided some evidence as to the likely cause of mortality – starvation, parasitism, and so on. It is therefore surprising that our estimates of the total nest-loss component should be smaller for each of the seven species than were the estimates of Snow & Mayer-Gross (1967), the differences in several cases being very substantial. This is in fact the result of a secular improvement in breeding success of these species: our data relate to the years 1962–80 whilst the earlier study was based on material for the years 1960–63, since when breeding success has risen in many species (Chapter 9).

Appendix 3: Sources and use of agricultural statistics

The agricultural statistics quoted are drawn from three main sources: (1) the June Census of MAFF and DAFS (2) the Surveys of Fertiliser Practice organised from Rothamsted Experimental Station, and (3) the Pesticide Usage Surveys organised by MAFF. Additional surveys of the numbers of machines on farms, which are conducted less frequently than annually, and surveys of farm types have also been used for these specific subjects.

The June Census

The June Census was instituted in 1866 and takes the form of a questionnaire sent to all occupiers of farm holdings. Completing and filing the return correctly is a statutory requirement. The questions listed cover the area of the holding and its basic characteristics, i.e. the area of crops and fallow (tillage), grassland, rough grazings, woodland and other land, followed by detailed questions on the area of each crop and of grass leys, the number of farmers and workers and of each type of farm stock – cattle, sheep, pigs and poultry and, formerly, horses. The questions for farm stock are further broken down into categories, for example to separate dairy cows from beef cows. Since 1975 horticultural crops and glass have been covered by a separate return.

These basic questions have been recorded annually since 1866 (1892 for rough grazings) but certain other questions are also covered at intervals, such as grain storage or artificially dried grass. Over time new headings have been included and old ones dropped, usually but not always in response to changes in farming practice. Here a significant loss affecting our attempt to analyse the impact of farming on birds has been that the age of leys and the area of grass mown are no longer recorded. Some information on this subject can still be drawn from the Surveys of Fertiliser Practice but is less satisfactory.

The results are analysed on a parish basis and published on a county basis. In the returns from the early 1960s to 1980 counties were grouped into regions on a broad agricultural or climatic basis. These regions were – *East*: Bedfordshire, Cambridgeshire, Essex, Greater London (part), Hertfordshire, Huntingdonshire, Lincolnshire (part), Norfolk, Suffolk; *South-east*: Berkshire, Buckinghamshire, Greater London (part), Hampshire and Wight, Kent, Oxfordshire, Surrey, Sussex; *East Midlands*: Derbyshire, Leicestershire, Lincolnshire (part), Northamptonshire, Nottinghamshire; *West Midlands*: Cheshire, Hereford & Worcester, Shropshire, Staffordshire, Warwickshire, West Midlands; *South-west*: Avon, Cornwall, Devon, Dorset, Gloucestershire, Scillies, Somerset, Wiltshire; *North*: Cleveland, Cumbria, Durham, North Yorkshire (part), Northumberland, Tyne & Wear; *Yorkshire & Lancashire*: Greater Manchester, Humberside, Lancashire, Merseyside, North Yorkshire (part), South and West Yorkshire; *Wales*: all counties.

We have used these regions as the basis of our maps, superimposing them on each historical period analysed, to give a clear and simple picture. In doing so the effects of county boundary changes were ignored. The only region where these had a major impact in 1974 was Yorkshire & Lancashire, where the creation of Humberside caused *c.* 80000 ha of Lincolnshire to shift into the region from the East Midland region. This change had the effect of varying the proportions of tilled land and pasture in each region by only 2–3 per cent. In Scotland the pre-1974 DAFS regions were similarly retained, despite the anomalous treatment of Caithness which resulted.

In a study of the Census statistics up to 1966 (HMSO, 1968), the Ministry noted that recording in any category becomes more accurate over time and that historical comparisons based on a single year are liable to be distorted by particular weather patterns in that year. We have therefore based historical comparisons on an average over three years, which is long enough to meet these difficulties while avoiding the additional distortion now possible from the speed with which changes occur in farming.

The county basis of recording in the Census makes a convenient unit to compare analyses of the CBC and Nest Record Cards with types of farming. To do this, counties were grouped as cereal counties, tillage counties, etc., according to the basic characteristics of their farming as shown in the Census. The groupings used and their definitions are listed below. In each of the groupings selected the characteristics used have remained constant over the period of the CBC. Counties shown in brackets no longer exist as administrative units but the land area remains in the grouping shown.

Tillage counties

These counties have 70 per cent of total land area as tilled land and at least 30 per cent of tilled land in root, vegetable or crops other than cereals and oilseed rape; they therefore have a high proportion of spring tillage.

Cambridgeshire	Humberside	Lincolnshire
(East Riding)	(Huntingdonshire)	Norfolk

Cereal counties

These counties have twice as much tilled land as pasture and a minimum of 80 per cent of tilled land in cereals. Root and vegetable crops total under 10 per cent of tilled land and in the group as a whole.

Essex	Hampshire	Northamptonshire
Hertfordshire	Oxfordshire	Warwickshire
Berkshire	Nottinghamshire	Wiltshire

Sheep counties

These counties have over one million sheep and twice as much pasture as tilled land. Between them they account for 55 per cent of the total sheep in England and Wales.

Devon	Northumberland
Cumbria (Cumberland and Westmorland)	All Welsh counties

Pasture counties

These counties have about 40 per cent of their area as pasture. Devon is common to this list and that for sheep and the two groups probably share many characteristics.

Cornwall	Devon	Somerset
Derbyshire	Lancashire	

Reduction of pasture

The following counties are the only ones in Britain in which the area of pasture has halved over the period of the CBC.

Berkshire	Lincolnshire
Essex	Humberside (formerly East Riding)

The following groups of counties were selected to examine changes in cereal management, particularly the change in the timing of cultivations from spring-sown to autumn-sown crops.

Wheat counties

The following counties are the major wheat-growing counties of England, between them accounting for 40–50 per cent of the wheat crop. Over the period of the CBC wheat has increased by *c*. 90 per cent in area in England and Wales, from 27 per cent of the cereals area to 47 per cent of a cereals area which has itself increased by a fifth. Only 1.5 per cent of the crop is now planted in spring compared to 13 per cent in the early 1960s.

Cambridgeshire	Leicestershire	Northamptonshire
Essex	Lincolnshire	Nottinghamshire
(Huntingdonshire)	Norfolk	Suffolk

Cereal-dominated counties

In these counties there is 50 per cent more tillage than pasture and at least 80 per cent of tillage is in cereals; 60 per cent of cereals is now autumn sown. Root and vegetable crops are not widely grown.

Hertfordshire	Gloucestershire	Oxfordshire
Buckinghamshire	Leicestershire	Warwickshire
Bedfordshire	Northamptonshire	Wiltshire
		West Sussex

Barley-dominated counties

Barley cannot be very clearly separated from cereals generally but the following counties have 50 per cent or more of tillage in barley and 50 per cent of barley is now autumn sown there.

Berkshire	Hampshire
Buckinghamshire	Oxfordshire
Gloucestershire	Wiltshire

Distribution of cattle and sheep

In five counties – Avon, Cheshire, Dorset, Hertfordshire and Suffolk – there are at least twice as many cattle as sheep per 1000 ha of total county area. Substantial grasslands are therefore more likely to carry cattle than sheep. The counties of East Sussex, Kent, Hereford & Worcester, Northamptonshire, Shropshire and Warwickshire show the opposite characteristic. In lowland England it is also possible to divide MAFF regions into counties with high cattle numbers and those with low. In the high category shown below, cattle numbers are 50 per cent higher per 1000 ha of land area than in the low category.

Region	*High cattle*	*Low cattle*
East	Hertfordshire, Norfolk, Suffolk	Cambridgeshire, Essex, Bedfordshire
South-east	Buckinghamshire, Sussex, Oxfordshire	Hampshire, Kent, Surrey

East Midlands	Derbyshire, Leicestershire	Lincolnshire, Northamptonshire, Nottinghamshire
West Midlands	Cheshire, Staffordshire	Hereford & Worcester, Shropshire, Warwickshire
South-west	Cornwall, Dorset, Somerset	Devon, Gloucestershire, Wiltshire

The concept of stocking rates for sheep and cattle is widely used throughout this book. These are expressed as grazing units calculated on the basis of 1 head of cattle = 5 grazing units; 1 sheep = 1 grazing unit.

The stocking rates are calculated on the assumption that these animals are kept solely by the grass area. This is not, of course, true, as arable crops have long been fed to cattle and sheep, a management practice which has also changed over time. Grass, however, remains as it has always been, the basic resource used by such farm stock. Thus stocking rates, as defined here, are used as a simple measure of increasing or decreasing pressure on grassland habitat. In this the use of arable crops as stock feed is irrelevant.

Surveys of Fertiliser Practice

These surveys are carried out annually and are voluntary. They are based on a stratified random sample of farms, each of which is visited by a fieldworker who consults the farmer to complete a detailed questionnaire on the fertiliser practice for each crop grown on the farm. Samples within the strata are chosen so that the number of farms surveyed in any region is representative of the proportion of crops and grass within that region (Church & Leech, 1983). The results are published on a basis of crops or grassland use and have been analysed on that basis. But information is also given on a regional basis, using four farm type regions – arable, arable/dairy, dairy and livestock.

The surveys also include much valuable incidental information on basic farm management patterns, particularly in grassland. The map of farm type regions was therefore laid over that of MAFF regions to make estimates of the regional extent of management practices, such as the proportion of grass cut for hay or silage.

Pesticide Usage Surveys

Again these surveys are voluntary. Because of the complexity of agrochemical usage however, the total range of crops examined by the surveys can be covered only over a four-year cycle. Thus survey work is conducted every year but each crop is surveyed only every fourth year. The information for the surveys is gathered by the same methods as the Surveys of Fertiliser Practice and published on a crop basis; the most recent surveys also include regional information based on the MAFF regions.

We have used the surveys for four crop groups – orchards, vegetables, roots and cereals – as these cover the bulk of the area of farmland outside grassland and enable a fairly simple picture of a complicated subject to be presented. Grassland was excluded because intensive management, where chemicals would be most frequently used, nearly always involves ploughing and reseeding and this habitat change cannot then be readily separated from the subsequent management input. In our tables vegetables include brassicas for human consumption and roots include potatoes, sugarbeet, roots and brassicas for stock feed and oilseed rape. Because of this, we have had to group the results for some individual surveys, for example, those for potatoes and sugarbeet.

The basic information analysed comprised area grown, the area sprayed and tonnage used. The first is straightforward. The area sprayed, however, is based on the frequency of spraying. In other words, if a field of 20 ha is sprayed three times, the area sprayed is 60 spray hectares and is recorded as 60 ha under 'area sprayed'. The tonnage used is of the active chemical ingredient in every case, not products.

Although the surveys are not analysed on a regional basis, some examination of regional trends was made, for example, for organochlorine usage. To do this the assumption was made that chemical usage was uniformly distributed over the national crop area. This is not always so but the marked regional concentrations of many crops does much to circumvent this problem. The overall result is to flatten differences between regional levels of use but the broad regional patterns and comparisons over time remain valid.

The other source of information used for agrochemicals is the ACAS Approved List. This is published annually by MAFF and comprises a list of insecticides, fungicides and herbicides approved under the Agricultural Chemicals Approval Scheme for use on farm crops. The Lists are not exhaustive as they omit some chemicals cleared for farm use under the Pesticides Safety Precaution Scheme. The criteria adopted for approval under the ACAS scheme are efficiency and safety. Despite being incomplete, however, the ACAS Approved Lists provide the best consistent information on general trends and patterns in chemical usage and methods of application over the 30 year period of the agrochemical revolution. The number of chemicals in each main category have been analysed and used as an indicator of the changing scale of chemical application.

Appendix 4: Scientific names of birds mentioned in the text

Mute Swan *Cygnus olor*
Pink-footed Goose *Anser brachyrhynchus*
White-fronted Goose *A. albifrons*
Greylag Goose *A. anser*
Canada Goose *Branta canadensis*
Barnacle Goose *B. leucopsis*
Brent Goose *B. bernicla*
Wigeon *Anas penelope*
Mallard *A. platyrhynchos*
Kite *Milvus milvus*
Hen Harrier *Circus cyaneus*
Goshawk *Accipter gentilis*
Sparrowhawk *A. nisus*
Buzzard *Buteo buteo*
Golden Eagle *Aquila chrysaetos*
Kestrel *Falco tinnunculus*
Merlin *F. columbarius*
Peregrine *F. peregrinus*
Red-legged Partridge *Alectoris rufa*
Grey Partridge *Perdix perdix*
Pheasant *Phasianus colchicus*
Corncrake *Crex crex*
Moorhen *Gallinula chloropus*
Coot *Fulica atra*
Great Bustard *Otis tarda*
Oystercatcher *Haematopus ostralegus*
Stone Curlew *Burhinus oedicnemus*
Golden Plover *Pluvialis apricaria*
Lapwing *Vanellus vanellus*
Dunlin *Calidris alpina*
Snipe *Gallinago gallinago*
Black-tailed Godwit *Limosa limosa*
Curlew *Numenius arquata*

Redshank *Tringa totanus*
Black-headed Gull *Larus ridibundus*
Common Gull *L. canus*
Lesser Black-backed Gull *L. fuscus*
Herring Gull *L. argentatus*
Stock Dove *Columba oenas*
Woodpigeon *C. palumbus*
Collared Dove *Streptopelia decaocto*
Turtle Dove *S. turtur*
Cuckoo *Cuculus canorus*
Barn Owl *Tyto alba*
Little Owl *Athene noctua*
Tawny Owl *Strix aluco*
Long-eared Owl *Asio otus*
Short-eared Owl *A. flammeus*
Swift *Apus apus*
Kingfisher *Alcedo atthis*
Wryneck *Jynx torquilla*
Great Spotted Woodpecker *Dendrocopus major*
Crested Lark *Galerida cristata*
Skylark *Alauda arvensis*
Swallow *Hirundo rustica*
House Martin *Delichon urbica*
Meadow Pipit *Anthus pratensis*
Yellow Wagtail *Motacilla flava*
Pied Wagtail *M. alba*
Wren *Troglodytes troglodytes*
Dunnock *Prunella modularis*
Robin *Erithacus rubecula*
Redstart *Phoenicurus phoenicurus*
Whincat *Saxicola rubetra*
Stonechat *S. torquata*
Wheatear *Oenanthe oenanthe*
Blackbird *Turdus merula*
Fieldfare *T. pilaris*
Song Thrush *T. philomelos*
Redwing *T. iliacus*
Mistle Thrush *T. viscivorus*
Sedge Warbler *Acrocephalus schoenobaenus*
Reed Warbler *A. scirpaceus*

Dartford Warbler *Sylvia undata*
Lesser Whitethroat *S. curruca*
Whitethroat *S. communis*
Garden Warbler *S. borin*
Blackcap *S. atricapilla*
Chiffchaff *Phylloscopus collybita*
Willow Warbler *P. trochilus*
Goldcrest *Regulus regulus*
Spotted Flycatcher *Muscipapa striata*
Long-tailed Tit *Aegithalos caudatus*
Blue Tit *Parus caeruleus*
Great Tit *P. major*
Nuthatch *Sitta europaea*
Magpie *Pica pica*
Chough *Pyrrhocorax pyrrhocorax*
Jackdaw *Corvus monedula*
Rook *C. frugilegus*
Carrion Crow *C. corone*
Raven *C. corax*
Starling *Sturnus vulgaris*
House Sparrow *Passer domesticus*
Tree Sparrow *P. montanus*
Chaffinch *Fringilla coelebs*
Brambling *F. montifringilla*
Greenfinch *Carduelis chloris*
Goldfinch *C. carduelus*
Linnet *C. cannabina*
Twite *C. flavirostris*
Redpoll *C. flammea*
Bullfinch *Pyrrhula pyrrhula*
Yellowhammer *Emberiza citrinella*
Reed Bunting *E. schoeniclus*
Corn Bunting *Miliaria calandra*

Appendix 5: Scientific names of other animals mentioned in the text

Common shrew *Sorex araneus*
Pigmy shrew *S. minutus*
Water shrew *Neomys fodiens*
Rabbit *Oryctolagus cuniculus*
Bank vole *Clethrionomys glareolus*
Short-tailed vole *Microtus agrestis*
Wood mouse *Apodemus sylvaticus*
Harvest mouse *Micromys minutus*
House mouse *Mus musculus*
Brown rat *Rattus norvegicus*
Fox *Vulpes vulpes*
Stoat *Mustela erminea*

References

Alexander, W. B. (1932). The bird population of an Oxfordshire farm. *Journal of Animal Ecology*, **1**, 58–64.

Alexander, W. B. & Lack, D. (1944). Changes in status among British breeding birds. *British Birds*, **38**, 42–5, 62–9, 82–8.

Arnold, G. R. (1983). The influence of ditch and hedgerow structure, length of hedgerows, and area of woodland and garden on bird numbers on farmland. *Journal of Applied Ecology*, **20**, 731–50.

Bailey, S., Bunyan, P. J., Hamilton, G. A., Jennings, D. M. & Stanley, P. I. (1972). Accidental poisoning of wild geese in Perthshire, November 1971. *Wildfowl*, **23**, 88–91.

Baird, W. N. & Tarrant, J. R. (1973). *Hedgerow Destruction in Norfolk 1946–1970*. Norwich: Centre for East Anglian Studies, University of East Anglia.

Balen, J. H. van (1980). Population fluctuations of the Great Tit and feeding conditions in winter. *Ardea*, **68**, 143–64.

Barber, D. (Ed.). (1970). *Farming and Wildlife: A study in Compromise*. Sandy: Royal Society for the Protection of Birds.

Batten, L. A. (1977). *Studies on the population dynamics and energetics of Blackbirds, Turdus merula Linnaeus*. Ph.D. thesis, University of London.

Beintema, A. J. (1982). Meadow birds in The Netherlands. *Rijksinstituut voor Natuurbeheer – Rapport (Research Institute for Nature Management Annual Report) 1981*, 86–93.

Beintema, A. J., Beintema-Hietbrink, R. J. & Muskens, G. J. D. M. (1985). A shift in the timing of breeding in meadow birds. *Ardea*, **73**, 83–9.

Beintema, A. J., de Boer, T. F., Buker, J. B., Muskens, G. J. D. M., van der Wal, R. J. & Zegers, P. M. (1982). *Verstoring van weidevogellegsels door wedend vee*. Leersum: Rijksinstituut voor Natuurbeheer.

Benson, G. B. G. & Williamson, K. (1972). Breeding birds of a mixed farm in Suffolk. *Bird Study*, **18**, 34–50.

Beresford, M. (1975). *We Plough the Fields: Agriculture in Britain Today*. Harmondsworth: Penguin Books.

Bibby, C. J. (1978). Some breeding statistics of Reed and Sedge Warblers. *Bird Study*, **25**, 207–22.

Bowden, J. & Dean, G. J. W. (1977). The distribution of flying insects in and near a tall hedgerow. *Journal of Applied Ecology*, **14**, 343–54.

Boyce, D. V. H. (1979). The influence of sowing depth on the removal of grain from winter wheat by Starlings (*Sturnus vulgaris* L.). *Journal of Plant Pathology*, **28**, 68–71.

Boyd, H. (1980). Waterfowl crop damage prevention and compensation programs in the Canadian Prairie Provinces. In *Bird Problems in Agriculture*, ed. E. N. Wright, I. R. Inglis, & C. J. Feare, pp. 20–7. Croydon: British Crop Protection Council.

Brenchley, A. (1984). The use of birds as indicators of change in agriculture. In *Agriculture and the Environment*, ed. D. Jenkins, pp. 123–8. Cambridge: Institute of Terrestrial Ecology.

Broad, P. D. (1952). The occurrence of weed seeds in samples submitted for testing by the Official Seed Testing Station. *Journal of the National Institute for Agricultural Botany*, **6**, 275–86.

Bull, A. L., Mead, C. J. & Williamson, K. (1976). Bird-life on a Norfolk farm in relation to agricultural changes. *Bird Study*, **23**, 203–18.

Bunn, D. S., Warburton, A. B. & Wilson, R. D. S. (1983). *The Barn Owl*. Calton: Poyser.

Bunyan, P. J. & Stanley, P. I. (1983). The environmental cost of pesticide usage in the United Kingdom. *Agricultural Ecosystems and Environment*, **9**, 187–209.

Burton, P. & Osborne, P. (1980). Woodman – spare that tree! *Birds*, **8**, 35–7.

Buxton, J. (Ed.) (1981). *The Birds of Wiltshire*. Trowbridge: Wiltshire Library & Museum Service.

Cadbury, C. J. (1980a). The status and habitats of the Corncrake in Britain 1978–79. *Bird Study*, **27**, 203–18.

Cadbury, C. J. (1980b). *Silent Death: the destruction of birds and mammals through the deliberate misuse of poisons in Britain*. Sandy: Royal Society for the Protection of Birds.

Cadbury, C. J. (1984). The effects of flood alleviation and land drainage on birds of wet grasslands. In *Agriculture and the Environment*, ed. D. Jenkins, pp. 108–16. Cambridge: Institute of Terrestrial Ecology.

Campbell, B. (1953). A comparison of bird populations upon 'industrial' and 'rural' farmland in South Wales. *Transactions of the Cardiff Naturalists Society*, **81**, 4–65.

Cawthorne, R. A. & Marchant, J. H. (1980). The effects of the 1978/79 winter on British bird populations. *Bird Study*, **27**, 163–72.

Chancellor, R. J., Fryer, J. D. & Cussans, G. W. (1984). The effects of agricultural practices on weeds in arable land. In *Agriculture and the Environment*, ed. D. Jenkins, pp. 89–94. Cambridge: Institute of Terrestrial Ecology.

Chapman, W. M. M. (1939). The bird population of an Oxfordshire farm. *Journal of Animal Ecology*, **8**, 286–99.

Charles, J. K. (1972). *Territorial behaviour and the limitation of population size in Crows*, Corvus corone *and* C. cornix. Ph.D. thesis, University of Aberdeen.

Church, B. M. & Leech, P. K. (1983). *Fertiliser Use on Farm Crops in England and Wales*. London: MAFF.

Collinge, W. E. (1924–27). *The Food of some British Wild Birds*. York: Collinge.

Colquhoun, M. K. (1951). *The Woodpigeon in Britain*. London: HMSO.

Cooke, A. S., Bell A. A. & Haas, M. B. (1982). *Predatory Birds, Pesticides and Pollution*. Cambridge: Institute of Terrestrial Ecology.

Cox, S. (1984). *A New Guide to the Birds of Essex*. Essex Bird Watching and Preservation Society.

Cramp, S. (1963). Toxic chemicals and birds of prey. *British Birds*, **56**, 124–39.

Cramp, S. & Conder, P. J. (1961). *The Deaths of Birds and Mammals Connected with Toxic Chemicals*. Report No. 1 of the BTO–RSPB Toxic Chemicals Committee.

Cramp, S., Conder, P. J. & Ash, J. S. (1962). *Deaths of Birds and Mammals from Toxic Chemicals*. Report No. 2 of the Joint Committee of the BTO and RSPB in collaboration with the Game Research Association.

Cramp, S. & Simmons, K. E. L. (Eds) (1982). *The Birds of the Western Palearctic*. Oxford: Oxford University Press.

Davies, N. B. (1976). Food, flocking and territorial behaviour of the pied wagtail (*Motacilla alba yarrelli* Gould) in winter. *Journal of Animal Ecology*, **45**, 235–54.

Davies, N. B. & Lundberg, A. (1985). The influence of food on time budgets and timing of breeding of the Dunnock *Prunella modularis Ibis*, **127**, 100–10.

Davis, B. N. K. (1967). Bird feeding preferences among different crops in an area near Huntingdon. *Bird Study*, **14**, 227–37.

Dean, A. R. (1974). The rookeries of Warwickshire, Worcestershire and Staffordshire. *West Midlands Bird Report*, **37**, 14–18.

Deans, I. R. (1979). Feeding of brent geese on cereal fields in Essex and observations of the subsequent loss of yield. *Agro-Ecosystems*, **5**, 283–88.

Dobbs, A. (1964). Rook numbers in Nottinghamshire over 35 years. *British Birds*, **57**, 360–4.

Dobbs, A. (Ed.) (1975). *The Birds of Nottinghamshire*. London: David & Charles.

Dunnet, G. M. (1955). The breeding of the Starling *Sturnus vulgaris* in relation to its food supply. *Ibis*, **97**, 619–62.

Dunnet, G. M. & Patterson, I. J. (1968). The rook problem in north-east Scotland. In *The Problems of Birds as Pests*, ed. R. K. Murton & E. N. Wright, pp. 119–39. London: Academic Press.

Dunning, R. A. (1974). Bird damage to sugarbeet. *Annals of Applied Biology*, **76**, 325–66.

Dyer, M. I. & Ward, P. (1977). Management of pest species. In *Granivorous birds in Ecosystems*, ed. J. Pinowski & S. C. Kendeigh, pp. 267–300. Cambridge: Cambridge University Press.

Edwards, C. A. (1984). Changes in agricultural practice and their impact on soil organisms. In *Agriculture and the Environment*, ed. D. Jenkins, pp. 56–65. Cambridge: Institute of Terrestrial Ecology.

Edwards, P. J. (1977). 'Re-invasion' by some farmland bird species following capture and removal. *Polish Ecological Studies*, **3**, 53–70.

Elkington, J. (1978). A treeless Britain? *New Scientist*, **78**, 72–4.

Evans, P. R. (1968). Winter fat deposition and overnight survival of yellow buntings. *Journal of Animal Ecology*, **38**, 415–24.

Evans, P. (1972). The Common Birds Census: eight years at Ely. *Cambridge Bird Club Report*, **45**, 36–9.

Feare, C. J. (1974). Ecological studies of the Rook (*Corvus frugilegus* L.) in north-east Scotland. Damage and its control. *Journal of Applied Ecology*, **11**, 897–914.

Feare, C. J. (1978). The ecology of damage by Rooks (*Corvus frugilegus*). *Annals of Applied Biology*, **88**, 329–34.

Feare, C. J. (1980). The economics of starling damage. In *Bird Problems in Agriculture*, ed. E. N. Wright, I. R. Inglis & C. J. Feare, pp. 39–55. Croydon: British Crop Protection Council.

Feare, C. J. (1982). Birds and change on the farm. *The Living Countryside*, **155**, 3084–7.

Feare, C. J., Dunnet, G. M. & Patterson, I. J. (1974). Ecological studies of the Rook (*Corvus frugilegus* L.) in north-east Scotland: food intake and feeding behaviour. *Journal of Applied Ecology*, **11**, 867–96.

Feare, C. J. & Swannack, K. P. (1978). Starling damage and its prevention in a open-fronted calf yard. *Animal Production*, **26**, 259–65.

Flegg, J. J. M. (1980). Biological factors affecting control strategy. In *Bird Problems in Agriculture*, ed. E. N. Wright, I. R. Inglis & C. J. Feare, pp. 7–19. Croydon: British Crop Protection Council.

Fretwell, S. D. & Lucas, H. L. (1969). On territorial behaviour and other factors influencing habitat distribution. *Acta Biotheoretica*, **19**, 16–36.

Frost, R. A. (1978). *The Birds of Derbyshire*. Buxton: Moorland Publishing Company.

Fuchs, E. (1982). Folgen kulturtechnischer Massnahmen auf den Sommervogelbestand im schweizerischen Mittelland. *Ornithologische Beobachter*, **79**, 121–7.

Fuller, R. J. (1982). *Bird Habitats in Britain*. Calton: Poyser.

Fuller, R. J. (1984). *The distribution and feeding behaviour of breeding songbirds on cereal farmland at Manydown Farm, Hampshire, in 1984*. Report to the Game Conservancy. Tring: British Trust for Ornithology.

Fuller, R. J. & Glue, D. E. (1977). The breeding biology of the Stonechat and Whinchat. *Bird Study*, **24**, 215–28.

Fuller, R. J., Marchant, J. M. & Morgan, R. A. (1985). How representative of agricultural practice in Britain are Common Birds Census farmland plots? *Bird Study*, **32**, 60–74.

Fuller, R. J. & Youngman, R. E. (1979). The utilisation of farmland by Golden Plovers wintering in southern England. *Bird Study*, **26**, 37–46.

Galbraith, H. (1985). *The influence of habitat on the breeding success and behaviour of Lapwings* Vanellus vanellus *on farmland.* Research report 1985. Glasgow: University of Glasgow.

Galbraith, H. & Furness, R. W. (1983). Habitats and distribution of waders breeding on Scottish agricultural land. *Scottish Birds*, **4**, 98–107.

Gibb, J. (1954). Feeding ecology of tits, with notes on Treecreeper and Goldcrest. *Ibis*, **96**, 513–43.

Gilg, A. W. (1975). Agricultural land classification in Britain: a review of the Ministry of Agriculture's new Map Series. *Biological Conservation*, **7**, 74–77.

Gill, F. B. & Wolf, L. L. (1975). Economics of feeding territoriality in the golden-winged sunbird. *Ecology*, **56**, 333–45.

Glue, D. E. (1967). Prey taken by the Barn Owl in England and Wales. *Bird Study*, **14**, 169–83.

Glue, D. E. (1982). *The Garden Bird Book*. London: Macmillan.

Gooch, S. M. (1963). The occurrence of weed seeds in samples tested by the Official Seed Testing Station 1960–61. *Journal of the National Institute for Agricultural Botany*, **9**, 353–71.

Gordon, M. (1972). Reed Buntings on an Oxfordshire farm. *Bird Study*, **19**, 81–90.

Green, R. E. (1978). Factors affecting the diet of farmland skylarks *Alauda arvensis*. *Journal of Animal Ecology*, **47**, 913–28.

Green R. (1980). Food selection by skylarks: the effect of a pesticide on grazing preferences. In *Bird Problems in Agriculture*, ed. E. N. Wright, I. R. Inglis & C. J. Feare, pp. 180–87. Croydon: British Crop Protection Council.

Green, R. (1981). Double nesting in red-legged partridges. *Game Conservancy Annual Review for 1980*, **12**, 35–8.

Green, R. E. (1984). The feeding ecology and survival of partridge chicks (*Alectoris rufa* and *Perdix perdix*) on arable farmland in East Anglia. *Journal of Applied Ecology*, **21**, 817–30.

Greenhalgh, M. E. (1971). The breeding bird communities of Lancashire saltmarshes. *Bird Study*, **18**, 199–212.

Greenhalgh, M. E. (1973). A comparison of breeding success of the Oystercatcher between inland and coastal areas in N.W. England. *Naturalist*, **926**, 87–8.

Hamilton, G. A., Ruthven, D. A., Findlay, E., Hunter, K. & Lindsay, D. A. (1981). Wildlife deaths in Scotland resulting from misuse of agricultural chemicals. *Biological Conservation*, **21**, 315–26.

Hardy, A. R. & Stanley, P. I. (1984). The impact of the commercial agricultural use of organophosphorus and carbamate pesticides on British wildlife. In *Agriculture and the Environment*, ed. D. Jenkins, pp. 72–80. Cambridge: Institute of Terrestrial Ecology.

Harrison, G. R., Dean, A. R., Richards, A. J. & Smallshire, D. (1982). *The Birds of the West Midlands*. Studley: West Midland Bird Club.

Harthan, A. J. (1946). *The Birds of Worcestershire*. Worcester: Littlebury.

Helliwell, D. R. (1975). The distribution of woodland plant species in some Shropshire hedgerows. *Biological Conservation*, **7**, 61–72.

Hickling, R. (1978). *Birds in Leicestershire and Rutland*. Leicester: Leicester and Rutland Ornithological Society.

Hilden, O. & Koskimies, J. (1969). Effects of the severe weather of 1965/66 upon winter bird fauna in Finland. *Ornis Fennica*, **46**, 22–31.

HMSO (1968). *A Century of Agricultural Statistics*. London: HMSO.

Hogstedt, G. (1974). Length of the prelaying period in the Lapwing *Vanellus vanellus* in relation to its food resources. *Ornis Scandinavica*, **5**, 1–4.

Holyoak, D. (1968). A comparative study of the food of some British Corvidae. *Bird Study*, **15**, 147–53.

Holyoak, D. (1972). Food of the Rook in Britain. *Bird Study*, **19**, 59–68.

Holyoak, D. (1974). Territorial and feeding behaviour of the Magpie. *Bird Study*, **21**, 117–28.

Hooper, M. (1970). Hedges and birds. *Birds*, **3**, 114–17.

Hooper, M. D. (1984). What are the main recent impacts of agriculture on wildlife? Could they have been predicted, and what can be predicted for the future? In *Agriculture and the Environment*, ed. D. Jenkins, pp. 33–6. Cambridge: Institute of Terrestrial Ecology.

Horne, B. van (1983). Density as a misleading indicator of habitat quality. *Journal of Wildlife Management*, **47**, 893–901.

Hoskins, W. G. & Stamp, L. D. (1963). *The Common Lands of England and Wales*. London: Collins.

Hudson, R. (1965). The spread of the Collared Dove in Britain and Ireland. *British Birds*, **58**, 105–39.

Jackson, R. & Jackson, J. (1980). A study of Lapwing breeding population changes in the New Forest, Hampshire, *Bird Study*, **27**, 27–34.

Jepson, P. C. & Green, R. E. (1983). Prospects for improving control strategies for sugar-beet pests in England. *Advances in Applied Biology*, **7**, 175–250.

Johnson, D. H. (1979). Estimating nest success: the Mayfield method and an alternative. *Auk*, **96**, 651–61.

Jorgensen, O. H. (1975). (Breeding birds in farmland ponds of Djursland, eastern Jutland, 1973.) In Danish, with English summary. *Dansk Ornithologisk Forenings Tidsskrift*, **69**, 103–10.

Kent, A. K. (1964). The breeding habitats of the Reed Bunting and Yellowhammer in Nottinghamshire. *Bird Study*, **11**, 123–7.

Klett, A. T. & Johnson, D. H. (1982). Variability in nest survival rates and implications to nesting studies. *Auk*, **99**, 77–87.

Klomp, H. (1953). (*Habitat Selection in the Lapwing* Vanellus vanellus (*L.*).) In Dutch, with English summary. Leiden: E. J. Brill.

Kluijver, H. N. & Tinbergen, L. (1953). Territory and the regulation of density in titmice. *Archives Neerlandaises Zoologie*, **10**, 265–89.

Knight, A. C. & Shepherd, K. B. (1985). *The Effect of Agricultural Improvement on some Upland Breeding Birds in Wales*. Newtown: Royal Society for the Protection of Birds.

Krebs, J. R. (1971). Territory and breeding density in the Great Tit, *Parus major* L. *Ecology*, **52**, 1–22.

Krebs, J. R. (1977). Song and territory in the Great Tit. In *Evolutionary Ecology*, ed. B. Stonehouse & C. M. Perrins, pp. 47–62. London: Macmillan.

Lack, D. (1971). *Ecological Isolation in Birds*. Oxford: Blackwell Scientific Publications.

Lack, P. C. (1986). *The Atlas of Wintering Birds in Britain and Ireland*. Calton: Poyser (in press).

Laursen, K. (1980). (Bird censuses in Danish farmland, with an analysis of bird distributions in relation to some landscape elements.) In Danish, with English summary. *Dansk Ornithologisk Forenings Tidsskrift*, **74**, 11–26.

Laursen, K. (1981). Birds on roadside verges and the effect of mowing on frequency and distribution. *Biological Conservation*, **20**, 59–68.

Lister, M. D. (1964). The Lapwing habitat enquiry, 1960–61. *Bird Study*, **11**, 128–47.

Locke, G. M. (1962). A sample survey of field and other boundaries in Great Britain. *Quarterly Journal of Forestry*, **56**, 137–44.

Lomas, P. D. R. (1968). The decline of the Rook population of Derbyshire. *Bird Study*, **15**, 198–205.

Luder, R. (1983). Verteilung und Dichte der Bodenbruter im offenen Kulturland des Schweizererischen Mittellandes. *Ornithologishe Beobachter*, **80**, 127–32.

MacArthur, R. H. (1958). Population ecology of some warblers of north-eastern coniferous forests. *Ecology*, **39**, 599–619.

Macdonald, D. (1977). Predation on nestling Collared Doves. *Bird Study*, **24**, 126.

Macdonald, R. A. (1983). *The feeding and breeding ecology of the Rook* (Corvus frugilegus *L.*) *in Eastern Ireland*. Ph.D. thesis, University College, Dublin.

McGinn, D. B. & Clark, H. (1978). Some measurements of Swallow breeding biology in lowland Scotland. *Bird Study*, **25**, 109–18.

MAFF (1964). *Review of Persistent Organochlorine Pesticides: Report of the Advisory Committee on Poisonous Substances used in Agriculture and Food Storage*. London: MAFF.

Manley, G. (1952). *Climate and the British Scene*. London: Collins.

Marchant, J. H. (1980). Recent trends in Sparrowhawk numbers in Britain. *Bird Study*, **27**, 152–4.

Marchant, J. H. (1981). Residual edge effects with the mapping bird census method. *Studies in Avian Biology*, **6**, 488–91.

Marquiss, M., Newton, I. & Ratcliffe, D. A. (1978). The decline of the Raven, *Corvus corax*, in relation to afforestation in southern Scotland and northern England. *Journal of Applied Ecology*, **15**, 129–44.

Mason, C. F. (1976). Breeding biology of the *Sylvia* warblers. *Bird Study*, **23**, 213–32.

Mason, C. F. & Lyczynski, F. (1980). Breeding biology of the Pied and Yellow Wagtails. *Bird Study*, **27**, 1–10.

Mason, C. F. & Macdonald, S. M. (1976). Aspects of the breeding biology of the Snipe. *Bird Study*, **23**, 33–8.

Mayfield, H. (1961). Nesting success calculated from exposure. *Wilson Bulletin*, **73**, 255–61.

Mayfield, H. (1975). Suggestions for calculating nest success. *Wilson Bulletin*, **87**, 456–66.

Mellanby, K. (1981). *Farming and Wildlife*. London: Collins.

Middleton, A. D. & Huband, P. (1965). Increase in red-legged partridges (*Alectoris rufa*). *Game Research Association Annual Report*, **5**, 14–25.

Moles, R. (1975). *Wildlife diversity in relation to farming practice in County Down, N. Ireland*. Ph.D. thesis, Queen's University of Belfast.

Møller, A. P. (1980). (The impact of changes in agricultural use on the fauna of breeding birds: An example from Vendsyssel, North Jutland.) In Danish, with English summary. *Dansk Ornithologisk Forenings Tidsskrift*, **74**, 27–34.

Møller, A. P. (1983). Breeding habitat selection in the Swallow *Hirundo rustica*. *Bird Study*, **30**, 134–142.

Moore, N. W. (1962). Toxic chemicals and birds: the ecological background to conservation problems. *British Birds*, **55**, 428–36.

Moore, N. W. (1965). Pesticides and birds – a review of the situation in Great Britain. *Bird Study*, **12**, 222–52.

Moore, N. W. (1980). How many wild birds should farmland support? In *Bird Problems in Agriculture*, ed. E. N. Wright, I. R. Inglis & C. J. Feare, pp. 2–6. Croydon: British Crop Protection Council.

Moore, N. W. & Hooper, M. D. (1975). On the number of bird species in British woods. *Biological Conservation*, **8**, 239–50.

Moore, N. W., Hooper, M. D. & Davis, B. N. K. (1967). Hedges I. Introduction and reconnaissance studies. *Journal of Applied Ecology*, **4**, 201–20.

Morgan, R. A. (1982). *Breeding Seasons of some British Waders*. CST Commissioned Research Report Series. Huntingdon: Nature Conservancy Council.

Morgan, R. A. (1986). Changes in the breeding avifauna of agricultural land in lowland Britain. Moscow: Proceedings of the 18th International Ornithological Congress, 588–593.

Morgan, R. A. & O'Connor, R. J. (1980). Farmland habitat and Yellowhammer distribution in Britain. *Bird Study*, **27**, 155–62.

Morgan, R. A., Sage, B. & Vernon, R. (1979). 1980 Rookeries Survey. *BTO News*, **113**, 5.

Moss, D., Taylor, P. N. & Easterbee, N. (1979). The effects on songbird populations of upland afforestation with spruce. *Forestry*, **52**, 129–50.

Murfitt, R. C. & Weaver, D. J. (1983). Survey of the breeding waders of wet meadows in Norfolk. *Norfolk Bird and Mammal Report*, **26**, 196–201.

Murton, R. K. (1958). The breeding of Wood-pigeon populations. *Bird Study*, **5**, 157–83.

Murton, R. K. (1965). *The Wood-pigeon*. London: Collins.

Murton, R. K. (1966). Natural selection and the breeding season of the Stock Dove and Woodpigeon. *Bird Study*, **13**, 311–27.

Murton, R. K. (1968). Some predator-prey relationships in bird damage and population control. In *The Problems of Birds as Pests*, ed. R. K. Murton & E. N. Wright, pp. 157–69. London: Academic Press.

Murton, R. K. (1971). *Man and Birds*. Collins: London.

Murton, R. K., Isaacson, A. J. and Westwood, N. J. (1963). The food and growth of nestling wood-pigeons in relation to the breeding season. *Proceedings of the Zoological Society of London*, **141**, 747–82.

Murton, R. K., Isaacson, A. J. and Westwood, N. J. (1966). The relationships between wood-pigeons and their clover food supply and the mechanism of population control. *Journal of Applied Ecology*, **3**, 55–96.

Murton, R. K. & Westwood, N. J. (1974). Some effects of agricultural change on the English avifauna. *British Birds*, **67**, 41–69.

Murton, R. K. & Westwood, N. J. (1976). Birds as pests. *Applied Biology*, **1**, 89–181.

Murton, R. K., Westwood, N. J. and Isaacson, A. J. (1964). The feeding habits of the wood-pigeon *Columba palumbus*, stock dove *C. oenas* and turtle dove *Streptopelia turtur*. *Ibis*, **106**, 174–88.

Murton, R. K., Westwood, N. J. & Isaacson, A. J. (1974). A study of Wood-pigeon shooting: the exploitation of a natural animal population. *Journal of Applied Ecology*, **11**, 61–82.

Nature Conservancy Council (1977). *Nature Conservation and Agriculture*. London: Nature Conservancy Council.

Newton, A. (1896). *A Dictionary of Birds*. London: Black.

Newton, I. (1964a). The breeding biology of the Chaffinch. *Bird Study*, **11**, 47–68.

Newton, I. (1964b). Bud-eating by bullfinches in relation to the natural food supply. *Journal of Applied Ecology*, **1**, 265–79.

Newton, I. (1967). The adaptive radiation and feeding ecology of some British finches. *Ibis*, **109**, 33–98.

Newton, I. (1968). Bullfinches and fruit buds. In *The Problems of Birds as Pests*, ed. R. K. Murton & E. N. Wright, pp. 199–209. London: Academic Press.

Newton, I. (1970). Some aspects of the control of birds. *Bird Study*, **17**, 177–94.

Newton, I. (1972). *Finches*. London: Collins.

Newton, I. (1974). Changes attributed to pesticides in the nesting success of the sparrowhawk in Britain. *Journal of Applied Ecology*, **11**, 95–102.

Newton, I. (1979). *Population Ecology of Raptors*. Calton: Poyser.

Newton, I. (1980). The role of food in limiting bird numbers. *Ardea*, **68**, 11–30.

Newton, I. (1983). Birds and forestry. In *Forestry and Conservation*, ed. E. H. M. Harris, pp. 21–30. Tring: Royal Forestry Society.

Newton, I. & Haas, M. B. (1984). The return of the Sparrowhawk. *British Birds*, **77**, 47–70.

Newton, I. & Marquiss, M. (1976). Occupancy and success of Sparrowhawk territories. *Raptor Research*, **10**, 65–71.

Nicholson, E. M. (1938–39). Report on the Lapwing habitat enquiry, 1937. *British Birds*, **32**, 170–91; 207–29; 255–9.

O'Connor, R. J. (1980). Pattern and process in Great Tit (*Parus major*) populations in Britain. *Ardea*, **68**, 165–83.

O'Connor, R. J. (1984a). The importance of hedges to songbirds. In *Agriculture and the Environment*, ed. D. Jenkins, pp. 117–23. Cambridge: Institute of Terrestrial Ecology.

O'Connor, R. J. (1984b). *The Growth and Development of Birds*. Chichester: Wiley.

O'Connor, R. J. (1985). Behavioural regulation of bird populations: a review of habitat use in relation to migrancy and residency. In *Behavioural Ecology: ecological consequences of adaptive behaviour*, British Ecological Society Symposium 25, ed. R. M. Sibly & R. H. Smith, pp. 109–42. Oxford: Blackwell Scientific Publications.

O'Connor, R. J. (1986). Dynamical aspects of avian habitat use. In *Modeling Habitat Relationships in Terrestrial Vertebrates*, ed. J. Verner, M. L. Morrison & C. J. Ralph. Madison: University of Wisconsin Press (in press).

O'Connor, R. J. & Fuller, R. J. (1985). Bird population responses to habitat. In *Bird Census and Atlas Studies*, Proceedings of the 8th International Bird Census Conference, ed. K. Taylor, R. J. Fuller & P. C. Lack, pp. 197–211. Tring: British Trust for Ornithology.

O'Connor, R. J. & Marchant, J. H. (1979). *A field validation of some Common Birds Census techniques*. Nature Conservancy Council CST Commissioned Research Report Series. Huntingdon: Nature Conservancy Council.

O'Connor, R. J. & Mead, C. J. (1982). *Population Level and Nesting Biology of the Stock Dove* Columba oenas *in Great Britain, 1930–1980*. CST Commissioned Research Report Series. Huntingdon: Nature Conservancy Council.

O'Connor, R. J. & Mead, C. J. (1984). The Stock Dove in Britain, 1930–80. *British Birds*, **77**, 181–201.

O'Connor, R. J. & Shrubb, M. (1986). Some effects of agricultural development on British bird populations. In *Man and Birds*, Proceedings of the 'Man and Birds' Symposium, Johannesburg 1983, ed. L. J. Dunning. Johannesburg: Witwatersrand Bird Club (in press).

Orwin, C. S. & Whetham, E. H. (1964). *History of British Agriculture 1846–1964*. Newton Abbott: David & Charles.

Osborne, L. & Krebs, J. R. (1981). Replanting after Dutch elm disease. *New Scientist*, **90**, 212–15.

Osborne, P. (1982a). *The effects of Dutch elm disease on farmland bird populations*. D.Phil. thesis, Oxford University.

Osborne, P. (1982b). Some effects of Dutch elm disease on nesting farmland birds. *Bird Study*, **29**, 2–16.

Osborne, P. (1983). The influence of Dutch elm disease on bird population trends. *Bird Study*, **30**, 27–38.

Osborne, P. J. (1984). Bird numbers and habitat characteristics in farmland hedgerows. *Journal of Applied Ecology*, **21**, 63–82.

Osborne, P. & Osborne, L. (1980). The contribution of nest site characteristics to breeding-success among Blackbirds *Turdus merula*. *Ibis*, **112**, 512–17.

Owen, M. (1977). The role of wildfowl refuges on agricultural land in lessening the conflict between farmers and geese in Britain. *Biological Conservation*, **11**, 209–22.

Owen, M. (1980). The role of refuges in wildfowl management. In *Bird Problems in Agriculture*, ed. E. N. Wright, I. R. Inglis & C. J. Feare, pp. 144–56. Croydon: British Crop Protection Council.

Parr, D. (Ed.). (1972). *Birds in Surrey 1900–1970*. London: Batsford.

Parslow, J. L. F. (1968). Changes in status among breeding birds in Britain and Ireland. *British Birds*, **61**, 241–55.

Parslow, J. L. F. (1969). Breeding birds of hedges. *Monks Wood Experimental Station Report 1966–68*, 21.

Parslow, J. L. F. (1973). *Breeding Birds of Britain and Ireland*. Berkhamsted: Poyser.

Patterson, I. J. (1980). Territorial behaviour and the limitation of population density. *Ardea*, **68**, 53–62.

Peal, R. E. F. (1968). The distribution of the Wryneck in the British Isles 1964–1966. *Bird Study*, **15**, 111–26.

Perrins, C. M. (1970). The timing of birds' breeding seasons. *Ibis*, **112**, 242–55.

Peterson, R. T. & Fisher, J. (1956). *Wild America*. London: Collins.

Pettifor, R. A. (1983). Seasonal variation, and associated energetic implications, in the hunting behaviour of the Kestrel. *Bird Study*, **30**, 201–6.

Phillips, J. S. (1970). Inter-specific competition in Stonechat and Whinchat. *Bird Study*, **17**, 320–4.

Picozzi, N. & Weir, D. (1976). Dispersal and causes of death of Buzzards. *British Birds*, **69**, 193–201.

Pollard, E. Hooper, M. D. & Moore, N. W. (1974). *Hedges*. London: Collins.

Potts, G. R. (1970). Recent changes in the farmland fauna with special reference to the decline of the Grey Partridge. *Bird Study*, **17**, 145–66.

Potts, G. R. (1980). The effects of modern agriculture, nest predation and game management on the population ecology of partridges (*Perdix perdix* and *Alectoris rufa*). *Advances in Ecological Research*, **11**, 1–82.

Potts, G. R. (1981a). Insecticide sprays and the survival of partridge chicks. *Game Conservancy Animal Review*, **12**, 39–48.

Potts, G. R. (1981b). Fewer woodpigeons? *Game Conservancy Annual Review*, **12**, 83–7.

Potts, G. R. (1983). The Grey Partridge situation. *Game Conservancy Animal Review*, **14**, 24–30.

Potts, G. R. (1984a). Monitoring changes in the cereal ecosystem. In *Agriculture and the Environment*, ed. D. Jenkins, pp. 128–34. Cambridge: Institute of Terrestrial Ecology.

Potts, G. R. (1984b). Grey Partridges: how a computer model can help to solve practical game management questions. *Game Conservancy Annual Review*, **15**, 56–9.

Potts, G. R. & Vickerman, G. P. (1974). Studies on the cereal ecosystem. *Advances in Ecological Research*, **8**, 107–87.

Prater, A. J. (1982). Brent Goose feeding patterns for Chichester Harbour 1982/83. *Sussex Bird Report*, **35**, 87.

Prestt, I. (1965). An enquiry into the recent breeding status of some of the smaller birds of prey and crows in Britain. *Bird Study*, **12**, 196–221.

Prŷs-Jones, R. P. (1977). *Aspects of Reed Bunting ecology, with comparisons with the Yellowhammer*. D.Phil. thesis. Oxford University.

Prŷs-Jones, R. P. (1984). Migration patterns of the Reed Bunting, *Emberiza schoeniclus*, and the dependence of wintering distribution on environmental conditions. *Le Gerfaut*, **74**, 15–37.

Rands, M. (1982). The importance of nesting cover quality to partridges. *Game Conservancy Annual Review*. **13**, 58–64.

Rands, M. R. W. (1985). Pesticide use on cereals and the survival of grey partridge chicks: a field experiment. *Journal of Applied Ecology*, **22**, 49–54.

Ratcliffe, D. A. (1963). The status of the Peregrine in Great Britain. *Bird Study*, **10**, 56–90.

Ratcliffe, D. A. (1976). Observations on the breeding of the Golden Plover in Great Britain. *Bird Study*, **23**, 63–116.

Ratcliffe, D. A. (1977). *A Nature Conservation Review: the selection of biological sites of national importance to nature conservation in Britain*. (Two volumes). Cambridge: Cambridge University Press.

Ratcliffe, D. A. (1980). *The Peregrine Falcon*. Calton: Poyser.

Ratcliffe, D. A. (1984). The Peregrine breeding distribution in the United Kingdom in 1981. *Bird Study*, **31**, 1–18.

Redfern, C. P. F. (1982). Lapwing nest sites and chick mobility in relation to habitat. *Bird Study*, **29**, 201–8.

Ricklefs, R. E. (1969). An analysis of nesting mortality in birds. *Smithsonian Contributions to Zoology*, **9**, 1–48.

Robson, R. W. & Williamson, K. (1972). The breeding birds of a Westmorland farm. *Bird Study*, **19**, 202–14.

Rogers, J. G. (1980). Conditioned taste aversion: its role in bird damage control. In *Bird Problems in Agriculture*, ed. E. N. Wright, I. R. Inglis & C. J. Feare, pp. 173–9. Croydon: British Crop Protection Council.

Rose, L. N. (1982). Breeding ecology of British pipits and their Cuckoo parasite. *Bird Study*, **29**, 27–40.

RSPB (1984). *Hill Farming and Birds: a survival plan*. Sandy: Royal Society for the Protection of Birds.

Sage, B. L. & Vernon, J. D. R. (1978). The 1975 National Survey of Rookeries. *Bird Study*, **25**, 64–86.

Sage, B. L. & Whittington, P. A. (1985). The 1980 sample survey of rookeries. *Bird Study*, **32**, 77–81.

Salisbury, E. (1961). *Weeds and Aliens*. London: Collins.

Schoener, T. W. (1968). Size of feeding territories among birds. *Ecology*, **49**, 123–41.

Sharrock, J. T. R. (1976). *The Atlas of Breeding Birds in Britain and Ireland*. Tring: British Trust for Ornithology/Irish Wildbird Conservancy.

Shrubb, M. (1970). Birds and farming today. *Bird Study*, **17**, 123–44.

Shrubb, M. (1979). *The Birds of Sussex*. Chichester: Phillimore.

Shrubb, M. (1980). Farming influences on the food and hunting of Kestrels. *Bird Study*, **27**, 109–15.

Shrubb, M. (1985a). Breeding Sparrowhawks *Accipiter nisus* and organochlorine pesticides in Sussex and Kent. *Bird Study*, **32**, 155–63.

Shrubb, M. (1985b). Breeding habitats of the Lapwing in Sussex. *Sussex Bird Report*, **37**, 75–80.

Sitters, H. (1974). *Atlas of Breeding Birds in Devon*. Devon Bird Watching and Preservation Society.

Sly, J. M. A. (1977). *Review of usage of pesticides in agriculture and horticulture in England and Wales 1965–1974*. London: MAFF.

Sly, J. M. A. (1981). *Review of usage of pesticides in agriculture and horticulture in England and Wales 1975–1979*. London: MAFF.

Smith, W. (1949). *An Economic Geography of Great Britain*. London: Methuen.

Smith, K. W. (1983). The status and distribution of waders breeding on wet lowland grasslands in England and Wales. *Bird Study*, **30**, 177–92.

Snow, D. W. & Mayer-Gross, H. (1967). Farmland as a nesting habitat. *Bird Study*, **14**, 43–52.

Snow, B. K. & Snow, D. W. (1984). Long-term defence of fruits by Mistle Thrushes *Turdus viscivorus*. *Ibis*, **126**, 39–49.

Sotherton, N. (1984). The distribution and abundance of predatory arthropods overwintering in field boundaries. *Annals of Applied Biology*, **106**, 17–21.

Sotherton, N. (1985). An insecticidal action of some foliar fungicides. *Game Conservancy Annual Report*, **16**, 38–40.

Southwood, T. R. E. (1961). The number of species of insects associated with various trees. *Journal of Animal Ecology*, **30**, 1–8.

Southwood, T. R. E. & Cross, D. J. (1969). The ecology of the partridge. III. Breeding success and the abundance of insects in natural habitats. *Journal of Animal Ecology*, **38**, 497–509.

Spencer, R. (1982). Birds in winter – an outline. *Bird Study*, **29**, 169–82.

Spencer, R. (1983). Our changing avifauna. In *Enjoying Ornithology*, ed. R. A. O. Hickling, pp. 93–128. Calton: Poyser.

Stamp, L. D. (1955). *Man and the Land*. London: Collins.

Stanley, P. I. & Hardy, A. R. (1984). The environmental implications of current pesticide usage on cereals. In *Agriculture and the Environment*, ed. D. Jenkins, pp. 66–72. Cambridge: Institute of Terrestrial Ecology.

Stenger, J. (1958). Food habits and available food in relation to territory size. *Auk*, **75**, 335–46.

Stubbs, A. E. (1980). *Facts and Figures*. CST Notes 21. London: Nature Conservancy Council.
Sturrock, F. & Cathie, J. (1980). *Farm Modernisation and the Countryside*. Occasional Paper No. 12. Cambridge: University of Cambridge Department of Land Economy.
Swaine, C. M. (1982). *Birds of Gloucestershire*. Gloucester: Alan Sutton.
Tait, E. J. (1977). A method for comparing pesticide usage patterns between farmers. *Annals of Applied Biology*, **86**, 229–40.
Tapper, S. (1981). The effects of farming and Dutch elm disease on corvids. *Game Conservancy Annual Review for 1980*, **12**, 98–101.
Tapper, S. (1982). Predation and predator control tomorrow. *Game Conservancy Annual Review*, **13**, 72–80.
Taylor, K. (1984). The influence of watercourse management on Moorhen breeding biology. *British Birds*, **77**, 144–8.
Taylor, D. W., Davenport, D. L. & Flegg, J. J. M. (1984). *The Birds of Kent*. Rainham: Kent Ornithological Society.
Terrasson, F. & Tendron, G. (1981). The case for hedgerows. *Ecologist*, **11**, 210–21.
Ticehurst, C. B. (1932). *A History of the Birds of Suffolk*. London: Gurney & Jackson.
Tonkin, J. H. B. & Phillipson, A. (1973). The presence of weed seeds in cereal seed drills in England and Wales during spring 1970. *Journal of the National Institute for Agricultural Botany*, **13**, 1–8.
Tozer, D. & Taylor, I. R. (1979). *Farmland Hedges of East Lothian, 1978*. Edinburgh: Nature Conservancy Council, South East (Scotland) Region.
Trow-Smith, R. (1951). *English Husbandry*. London: Faber & Faber.
Verner, J. (1981). Measuring responses of avian communities to habitat manipulation. *Studies in Avian Biology*, **6**, 543–7.
Vernon, J. D. R. (1970). Feeding habits and food of the Black-headed and Common Gulls (Part 1 – Feeding habitats). *Bird Study*, **17**, 287–96.
Webber, M. I. (1975). *Some aspects of the non-breeding population dynamics of the Great Tit* (Parus major). D.Phil. thesis: Oxford University.
Wiens, J. A. & Johnston, R. F. (1977). Adaptive correlates of granivory in birds. In *Granivorous Birds in Ecosystems*, ed. J. Pinowski & S. C. Kendeigh, pp. 301–40. Cambridge: Cambridge University Press.
Wiens, J. A. & Dyer, M. I. (1977). Assessing the potential impact of granivorous birds in ecosystems. In *Granivorous Birds in Ecosystems*, ed. J. Pinowski & S. C. Kendeigh, pp. 205–66. Cambridge: Cambridge University Press.
Williamson, K. (1967). The bird community of farmland. *Bird Study*, **14**, 210–26.
Williamson, K. (1968). Buntings on a barley farm. *Bird Study*, **15**, 34–7.
Williamson, K. (1969). Habitat preferences of the Wren on English farmland. *Bird Study*, **16**, 53–9.
Williamson, K. (1970). Birds and modern forestry. *Bird Study*, **17**, 167–76.
Williamson, K. (1971). A bird census study of a Dorset dairy farm. *Bird Study*, **18**, 80–96.
Williamson, K. & Batten, L. A. (1977). Some ecological implications of the Common Birds Census. *Polish Ecological Studies*, **3**, 237–44.
Wright, E. N. (1973). Experiments to control Starling damage at intensive animal husbandry units. *European Plant Protection Organisation Bulletin*, **9**, 85–9.
Wright, E. N. (1980). General considerations. In *Bird Problems in Agriculture*, ed. E. N. Wright, I. R. Inglis & C. J. Feare, p.1. Croydon: British Crop Protection Council.
Wyllie, I. (1976). The bird community of an English parish. *Bird Study*, **23**, 39–50.
Zande, A. N. van der, Kewis, W. J. ter & Weijden, W. J. van der (1980). The impact of roads on the densities of four bird species in an open field habitat – evidence of a long distance effect. *Biological Conservation*, **18**, 299–321.

Glossary

ACAS Agricultural Chemicals Approval Scheme. This list, published annually by MAFF, provides farmers with a guide to the efficient and safe use of agrochemicals. It is not mandatory and does not include a number of chemicals which have PSPS approval.

ADAS Agricultural Development and Advisory Service – part of MAFF.

Arable land Land that is regularly ploughed or tilled although not necessarily annually. Arable land therefore includes leys (q.v.) and tillage (q.v.).

Brassicas Plants of the cabbage family Cruciferae. Several important arable field weeds, e.g. charlock *Sinapsis arvensis*, are brassicas, as are crops such as kale, rape, mustard and turnips.

Break crop A different crop inserted into a cereal rotation to break continuity, and vice versa.

CAP The Common Agricultural Policy of the EEC.

CBC The Common Birds Census of the British Trust for Ornithology (BTO). See Appendix 1.

Contact pesticide One that kills by direct contact.

Corn Cereals in Britain are universally known as corn, a term which in America applies specifically to maize.

DAFS The Department of Agriculture and Fisheries for Scotland.

Direct drilling Drilling into unploughed or uncultivated land.

Drill husbandry Sowing by using a drill rather than by broadcasting seed. The invention of the drill and planting crops in rows (which made weeding easier) was a major element in the agricultural revolution of the eighteenth century.

Drill to a stand Sowing crops such as sugarbeet or brassicas at the spacings at which they will be left to grow.

Drum mower A mower with two rotary cutters as distinct from a cutting bar with knife and fingers.

EEC The European Economic Community, also known as the Common Market.

Ensilage The process of making silage (q.v.)

Fallow Land that is ploughed or tilled but not sown. To till land without sowing, normally to control weeds.

Flash-harry A bird-scarer which alternately flashes fluorescent orange and black when turned by a breeze.

Forage harvester The machine used to gather mown or swathed grasses or other forage for silage.

FWAG The Farming and Wildlife Advisory Group.

Grazing units Since farm animals vary greatly in size, stocking rates (q.v.) are normally expressed in terms of uniform grazing units (one grazing unit = 1 sheep or 1/5th cow).

Growth regulator Chemicals now widely used to prevent lodging (q.v.) in cereals. They thicken the cells of the lower stem so making the plant shorter and stronger.

Hay Green fodder, normally grass and clover preserved by drying in sun and wind.

Headland The area at each edge of a field on which machines turn during working. The equivalent areas at the sides of a field are called *sidelands*.

Hedgerow, hedge Here treated as synonymous although some authors separate hedgerows as larger.

HGCA Home Grown Cereals Authority.

High farming A term normally taken to mean intensive rotation farming.

Hill farming Primarily livestock rearing and fattening, usually of sheep.

Holding The area of land farmed as a unit. It may include more than one farm.

Intervention A Government-operated market of last resort, designed to ensure minimum prices and therefore stabilise other markets.

Legumes Plants of the pea family Leguminosae which have the valuable capacity to fix atmospheric nitrogen in the soil.

Ley or temporary grass Grass or clover that is ploughed regularly in rotation.

Lodging Cereal crops flattened by rain or wind are said to have lodged. Certain diseases also cause or encourage lodging.

MAFF Ministry of Agriculture, Fisheries and Food.

Marginal land Land that is marginally profitable to farm; it is not necessarily inherently infertile.

Nitrogen Applied in the form of nitrates, the most widely used farm fertiliser.

Nurse crop A crop grown as cover for another, usually for a ley.

Organic farming Farming without inorganic fertilisers or pesticides.

Permanent grass Improved grass that is not ploughed regularly in rotation. Also known as pasture but that term should be strictly applied to grass that is grazed.

Pesticides Embraces the whole range of seed dressings, fungicides, herbicides, insecticides etc.

Pesticide Usage Surveys A regular cycle of surveys of the use of pesticides by farmers is conducted by MAFF every four years. The surveys are based on a stratified sample of farms. See Appendix 3.

Post-emergent herbicide One that is applied to crops that have emerged from the soil.

Precision drilling Sowing seeds at predetermined spacings and depths.

Pre-emergent herbicide A herbicide, which is applied to the crop immediately after sowing, before it emerges.

PSPS Pesticides Safety Precaution Scheme (see Preface).

Puddling The action of continuously treading wet ground which brings water to the surface. This then dries and forms a cap to the soil.

Reversible plough These ploughs are turned over at the end of each pass across the field, so eliminating ridges and furrows.

Root crops Strictly, plants in which the root is the commercially important part.

Rotations Cycles of crops grown regularly in succession.

Rough grazing Poor permanent grass. Rough grazing has no official definition and the line between it and permanent grass is always, therefore, a subjective opinion. Much of it in Britain today is hill or marsh.

Row crops A term applied to roots, brassica and vegetable crops which are grown in more widely spaced rows than other crops.

Self-binder The machine which cuts cereals and ties them into sheaves.

Semi-natural Used throughout to refer to those parts of the farmland environment that are not routinely managed for crop or grass production. This use includes but goes beyond the Nature Conservancy Council's definition of the term by including hedgerows, scrub, and similar habitat.

Sheepwalk Open unfenced hill or downland used for grazing.

Silage Green fodder preserved for winter keep in an airtight store or stack.

Stocking rate The number of animals kept per unit area of land; usually expressed in grazing units.

Systemic Of an insecticide, one that is absorbed by the plant to kill pests which subsequently prey on that plant.

Threshing drum The stationary machine used for threshing grain from sheaves.

Tillage crops Crops planted and harvested annually in tilled land.

Tilled land or tillage Land that is ploughed and cultivated annually. Thus it differs from arable which includes ley.

Tillering The process by which cereal or grass plants make multiple stems; each stem is known as a tiller.

Top dressing Applying fertiliser to the standing crop, as opposed to incorporating it in the seedbed.

Tramlines Permanent wheelings left in cereal crops along which go all passes of the sprayer and fertiliser distributor.

Upland Here used for land above 300 m.

Vegetables Horticultural crops for human consumption.

Volunteer (soldier) A self-sown plant (usually unwanted) of a previous crop.

Zero-grazing Cutting and carting grass to the cow rather than turning the cow out on the grass to graze.

Index